Food Additive User's Handbook

Food Additive User's Handbook

Edited by

JIM SMITH
Senior Food Scientist
Prince Edward Island Food Technology Centre
Canada

Blackie
Glasgow and London

Published in the USA by
avi, an imprint of
Van Nostrand Reinhold
New York

Blackie and Son Ltd
Bishopbriggs, Glasgow G64 2NZ
and
7 Leicester Place, London WC2H 7BP

Published in the United States of America by
AVI, an imprint of
Van Nostrand Reinhold
115 Fifth Avenue
New York, New York 10003

Distributed in Canada by
Nelson Canada
1120 Birchmount Road
Scarborough, Ontario M1K 5G4, Canada

16 15 14 13 12 11 10 9 8 7 6 5 4 3 2 1

© 1991 Blackie and Son Ltd
First published 1991

British Library Cataloging in Publication Data

Food additive user's handbook.
  1. Food. Additives.
  I. Jim Smith
  664.060212

  ISBN 0-216-92911-3

Library of Congress Cataloging-in-Publication Data

Food additive user's handbook / [edited] by James Smith.
      p.   cm.
    Includes bibliographical references and index.
    ISBN 0-442-31431-0
    1. Food additives.   I. Smith, James, Dr.
  TP455.F64   1991
  664'.06—dc20   91-6323
                                                  CIP

Phototypesetting by Interprint Ltd, Malta
Printed in Great Britain by Thomson Litho Limited, East Kilbride, Scotland

Preface

The aim of this book is to present technical information about the additives used in food product development, in a concise form. Food product development is an activity which requires application of technical skills and the use of a diverse range of information. Normally this information is scattered throughout the vast food science literature in journals and books and in technical publications from the various suppliers. It has been my experience, through consulting with the food industry, that there is a need for information on food additives in a quick-to-use form—in tables and figures where possible. Time wasted during information retrieval causes delay in practical development work, which results in delay of product launch and possibly the loss of market advantage.

This handbook will be used by food product development staff and by all food scientists requiring access to information on food additives in a quick-to-use format. Some knowledge of food science is assumed. Each chapter contains a bibliography which can be consulted if further information is required. Local legislation will have to be consulted to determine the legality of use of the additive, in which foods and at what level of addition. Information on safety can be found in *Food Additives Handbook* (1989) by R. J. Lewis, published by Van Nostrand Reinhold, New York.

The emphasis has been on practicality, so a strictly scientific presentation has not been followed. Also, due to differences in the nature of the various additives, there is variation in the proportion of tables to text. It is hoped that the user will find the presentation appropriate and useful. If there are any improvements, changes or additions you would like to suggest which would enhance the usefulness of the book, please let me know.

J. S.

How to use this book

This handbook is divided into chapters each of which covers a category of food additives with a specific function—antioxidants, flavours, colours and so on. The way in which the book is designed to be used is to first of all identify the *function* required of the additive. The appropriate chapter should then be consulted and the appropriate table or section examined for the additive which exhibits the necessary characteristics under the conditions which the food product imposes—pH, storage temperature, processing conditions etc.

Chapter 1 (antioxidants) begins with the nature of fats and oils and the problems associated with deterioration by hydrolysis and oxidation. The properties of the food-grade antioxidants are then addressed, followed by selection and performance. By consulting the section on performance, the additive to use in the desired situation may be determined.

Chapter 2 (sweeteners) first of all covers the properties of the various sweeteners and is followed by tables in which the properties of the sweeteners are summarised and through which sweeteners for specific applications may be chosen.

Chapter 3 (flavourings) has a useful glossary and concentrates on the application in the various food categories of flavourings by physical form. It also covers the dosage by application.

Chapter 4 (colours) covers first of all the properties of the various synthetic and natural colours in detail. The tables permit selection of colours for specific applications by presenting the properties and applications in an easy-to-read form. The chapter also summarises typical uses.

Chapter 5 (preservatives) begins with an introduction to the topic and includes tables of information on the properties of commonly permitted antimicrobial preservatives and other antimicrobial preservatives.

Chapter 6 (enzymes), after a brief introduction to enzymology, concentrates on the parameters which affect the choice of enzymes for food application, legislation and handling. Table 6.3 covers in detail the sources and characteristics of enzymes used in food processing and Table 6.4 lists the current applications of enzymes in food processing.

Chapter 7 (nutritive additives) starts with a short introductory text then continues with tables of information about the usage of fat-soluble vitamins (Table 7.1), water-soluble vitamins (Table 7.2), factors affecting the stability of vitamins added to foods (Table 7.3), overages of vitamins (Table 7.4), minerals (Table 7.5), RDAs (Tables 7.6 and 7.7), foods to which nutrients are commonly added (Table 7.8) and conversion factors (Table 7.9).

Chapter 8 (emulsifiers) contains an introduction and a glossary. It also includes classification, regulation and function, selection and the HLB concept and surfactants used in foods. The major applications of emulsifiers in foods are considered.

Chapter 9 (bulking agents) covers the nature of bulking agents and concentrates on traditional bulking agents, bulking agents as fat substitutes and extenders, and the utilisation of bulking agents.

Chapter 10 (pH control agents) covers the properties of common food acidulants and indicates the major natural acids in fruits and vegetables. A range of tables is presented which illustrates the application of ascorbic acid in a variety of situations.

Chapter 11 (hydrocolloids) begins with the classification, structure and applications of food hydrocolloids and then concentrates on the properties of hydrocolloids and indicates criteria for their selection.

Chapter 12 (antifoams and release agents) describes the properties of the various antifoams and selection of product type and grade.

Chapter 13 (flour improvers and raising agents) introduces flour improvers and raising agents and provides tables which illustrate the properties and criteria for selection of the additives.

Chapter 14 (gases) after an introduction covers freezing and chilling, gas atmospheres and special uses.

Chapter 15 (chelating agents) deals with chelating agents in foods, stability constants of metal chelates, chelating values, chemical and physical properties of chelating agents and applications in foods.

Contributors

Diediet Biebaut Puratos Corporation, Zone 1, Av. Industria 25, 1720 Groot-Bijgaarden, Belgium.

Dan F. Buck Eastman Chemical Products Inc., Kingsport, Tennessee, TE 37662, USA.

Luisita Dolfini Hoffman LaRoche Ltd., Dept. VM/H, CH-4002, Basle, Switzerland.

Robert J. Gordon Givaudan (Canada) Inc., 98 Walker Drive, Brampton, Ontario L6T 4H6, Canada.

Tom E. Luallen A. E. Staley, 2200 East Eldorado Street, Decatur, IL 62521-1801, USA.

Sue Marie Meat and Livestock Commission, P.O. Box 44, Winterhill House, Snowdon Drive, Milton Keynes, MK6 1AX, UK.

George R. McCain BOC Group Inc., 100 Mountain Avenue, Murray Hill, NJ 07974, USA.

Tjitte Nauta Akzo Chemicals Bv. Research Centre, Emmastraat 33, P.O. Box 10, 7400 A.A., Deventer, The Netherlands.

Yaw J. Owusu-Ansah POS Pilot Plant Corp., 118 Veterinary Road, Saskatoon, Saskatchewan S7N 2R4, Canada.

Doryne Peace Hoffman LaRoche Ltd., 401 The West Mall, Suite 700, Etobicoke, Ontario M9C 5J4, Canada.

Peter B. Rayner Overseal Foods, Park Road, Overseal, Burton-on-Trent, Staffordshire, DE12 6JX, UK.

Basil S. Kamel Atkemix Inc., P.O. Box 1085, Brantford, Ontario N3T 5T2, Canada.

Jim Smith   Prince Edward Island Food Technology Centre, P.O. Box 2000, Charlottetown, P.E.I. C1A 7N8, Canada.

Peter de la Teja   Food Products Takeda USA Inc., 8 Corporate Drive, Orangeburg, NY 10962-2614, USA.

Jens E. Trudso   Hercules Inc., 33 Sprague Avenue, Middletown, NY 10940, USA.

Anthony A. Zotto   G.E. Silicones, Waterford, NY 12188, USA.

Contents

3 Flavourings
R. J. GORDON

1 Antioxidants

D. F. BUCK

1.1 Introduction

For almost half a century, food-grade antioxidants have been routinely and intentionally added to food products to delay or inhibit free radical oxidation of fats and oils and the resulting off-odours and flavour known as rancidity. During this time, literally hundreds of compounds – both natural and synthetic – have been evaluated for antioxidant effectiveness and human safety. To date, approximately twenty compounds have met the stringent health, safety and performance parameters required by various food regulatory agencies to be approved as direct food additives (Table 1.1).

Table 1.1 Antioxidants approved for food use

Canada	United States	European Economic Community (EEC)
Ascorbic acid	BHA	Ascorbic acid
Ascorbyl palmitate	BHT	Ascorbyl palmitate
Ascorbyl stearate	Dilauryl thiopropionate	BHA
BHA	Glycine	BHT
BHT	Gum guaiac	Calcium ascorbate
Citric acid	4-Hydroxymethyl-2,6-	Dodecyl gallate
Gum guaiac	ditertiarybutyl phenol	Octyl gallate
Lecithin	Lecithin	Propyl gallate
Lecithin citrate	Propyl gallate	Sodium ascorbate
Monoglyceride citrate	TBHQ	Tocopherols
Monoisopropyl citrate	THBP (trihydroxybutyrophenone)	
Propyl gallate	Thiodipropionic acid	
Tartaric acid	Tocopherols	
Tocopherols		

Of these approved antioxidants, only five products find widespread usage throughout the world. These are:

- BHA (butylated hydroxyanisole)
- BHT (butylated hydroxytoluene)
- Propyl gallate
- TBHQ (tertiarybutyl hydroquinone)
- Tocopherols

These products, used either singly or in combination and often combined with acid synergists such as citric acid and ascorbic acid and their corresponding esters, fill the majority of the world's need for antioxidants in food products.

The primary objective of this chapter is to assist the food scientist in selecting and applying the most effective antioxidant for his or her intended use. This will be accomplished by demonstrating the relative effectiveness of various antioxidants in many of the world's fats and oils and also in popular food products and processes. Prior to this, the composition of fats and oils, the chemistry of fat oxidation, the mechanisms and functions of antioxidants and the individual properties of BHA, BHT, propyl gallate, TBHQ and tocopherols will be reviewed. In conclusion, some of the regulatory aspects of antioxidant usage will be addressed.

1.2 Fats and oils

Fats and oils are present in almost all foods. They provide important nutritional functions such as supplying caloric content to the food, serving as a source of essential fatty acids such as linoleic acid and solubilising vitamins A, D, E and K. Additionally, fats and oils provide important sensory properties to foods such as 'richness', desirable textural properties such as 'shortness' and a feeling of satisfaction when the food is consumed.

1.2.1 Composition

Chemically, fats and oils are triglycerides. They are esters of glycerol and long-chain fatty acids. In the chemical structure of triglycerides (1) R_1, R_2,

$$
\begin{array}{c}
\text{H} \quad\quad \text{O} \\
\text{|} \quad\quad\quad \text{||} \\
\text{H—C—O—C—R}_1 \\
\text{|} \quad\quad\quad \text{O} \\
\quad\quad\quad\quad \text{||} \\
\text{H—C—O—C—R}_2 \\
\text{|} \quad\quad\quad \text{O} \\
\quad\quad\quad\quad \text{||} \\
\text{H—C—O—C—R}_3 \\
\text{|} \\
\text{H}
\end{array}
$$

(1)

and R_3 represent the many fatty acids that can be esterified with glycerol. Most fatty acids found in food fats and oils contain 12–18 carbons and are chemically either saturated (containing carbon single bonds —C—C—) or unsaturated (containing carbon double bonds —C=C—). The chain length, degree of unsaturation and position of the fatty acid group when esterified with glycerol determine the physical and chemical characteristics of fats and oils.

1.2.2 Deterioration of fats and oils

Deterioration of food fats and oils primarily occurs from two chemical reactions – hydrolysis and oxidation. Hydrolysis of fats and oils occurs when water reacts with the triglyceride to form glycerol, mono- and diglycerides, and free fatty acids. Hydrolysis of crude fats and oils – both animal and vegetable – is often catalysed by lipase enzymes and results in substantial refining losses. These enzymes are deactivated by the thermal conditions of rendering and/or refining.

Hydrolysis. Hydrolysis of fats and oils is most prevalent in high-temperature food processing such as deep-fat frying of foods with a high water content (e.g. potatoes). Hydrolysis of fats and oils results in:

(1) lowered smoke point;
(2) foaming during frying;
(3) corrosion of food processing equipment due to free fatty acids;
(4) bitter or soapy flavours due to free fatty acids (especially when using coconut or palm kernel oil) – sometimes referred to as hydrolytic rancidity.

It is important to note that neither antioxidants nor other food additives can alleviate problems associated with hydrolysis of lipids. Hydrolysis is best controlled by using high-quality ingredients and good manufacturing practices.

Oxidation. This is the most common problem associated with the production, storage and usage of food fats and oils. Oxidation and subsequent rancidity of the lipids in many finished food products are often limiting factors in the food's shelf-life. Oxidation of unsaturated fats and oils is initiated with the formation of free radicals on exposure of the lipid to heat, light, metal ions and oxygen. This reaction occurs on the methylene groups adjacent to the carbon–carbon double bond (2). The reaction proceeds when the fat-free radical reacts with oxygen to form peroxide-free

$$\left[\cdots C - \underset{\underset{H}{|}}{\overset{\overset{H}{|}}{C}} - C = C - C \cdots\right] \xrightarrow{-H\bullet} \left[\cdots C - \underset{\bullet}{\overset{\overset{H}{|}}{C}} - C = C - C \cdots\right]$$

Fat   Fat Free Radical

(2)

$$\left[\cdots C-\overset{\overset{\displaystyle H}{|}}{\underset{\displaystyle \bullet}{C}}-C=C-C\cdots\right] + O_2 \rightarrow \left[\cdots C-\overset{\overset{\displaystyle H}{|}}{\underset{\overset{\displaystyle O}{\underset{\displaystyle \bullet}{|}}}{C}}-C=C-C\cdots\right]$$

Fat Free Radical

Peroxide Free Radical

(3)

radicals (3). The peroxide-free radical then extracts a hydrogen ion from another fat molecule forming a hydroperoxide and another fatty free radical (4). This is the propagating step of free radical oxidation making it a 'chain'

$$\left[\cdots C-\overset{\overset{\displaystyle H}{|}}{\underset{\overset{\displaystyle O}{\underset{\displaystyle \bullet}{|}}}{C}}-C=C-C\cdots\right] + \left[\cdots C-\overset{\overset{\displaystyle H}{|}}{\underset{\displaystyle H}{C}}-C=C-C\cdots\right] \rightarrow$$

Peroxide Free Radical

Fat

$$\left[\cdots C-\overset{\overset{\displaystyle H}{|}}{\underset{\overset{\displaystyle O}{\underset{\overset{\displaystyle O}{\underset{\displaystyle H}{|}}}{|}}}{C}}-C=C-C\cdots\right] + \left[\cdots C-\overset{\overset{\displaystyle H}{|}}{\underset{\displaystyle \bullet}{C}}-C=C-C\cdots\right]$$

Hydroperoxide

Fat Free Radical

(4)

type reaction. As a final step, the hydroperoxides decompose into smaller organic compounds such as aldehydes, ketones and acids. These decomposition products produce the rancid odours and flavour that characterise rancid fats and oils.

1.2.3 *Promoters of oxidation*

Recognition of catalysts, factors and conditions that promote or contribute to oxidation of lipids is essential if these conditions are to be controlled. Some of the most important factors are:

1. *Heat.* Oxidation is typical of most chemical reactions. A 10°C increase in temperature doubles the reaction rate.
2. *Light.* Ultraviolet light is a powerful initiator and catalyst of oxidation.

3. *Heavy metals.* Dissolved metals such as iron and copper are also effective catalysts for oxidation at the ppm level.
4. *Alkaline conditions.* Basic conditions and alkaline metal ions promote free radical oxidation.
5. *Degree of unsaturation.* Number and position of double bonds in fat molecules can be directly related to susceptibility to oxidation; for example, linolenic acid is more easily oxidised than oleic acid.
6. *Pigments.* Residual pigments found in vegetable oils promote oxidation. Chlorophyll, for example, can promote oxidation of various oils by producing singlet oxygen.
7. *Availability of oxygen.* Oxygen is required for oxidation.

1.2.4 *Recognising an oxidation problem*

Recognition of symptoms that allow the diagnosis of an oxidation problem is also important. Evidence that oxidation has occurred is generally quite apparent. Obvious signs are:

1. *Rancidity.* Oxidation of fats and oils results in off-odours and off-flavours.
2. *Colour changes.* Oxidation can result in either lightening or darkening in colour, depending upon the substrate. For example, fats and oils tend to darken. Pigments, especially carotenoids, tend to lighten.
3. *Loss of odour.* Essential oils such as citrus oils lose their distinctive fruity high notes of odour.
4. *Viscosity changes.* Fats and oils tend to thicken due to polymerisation that is catalysed by free radicals.

Less obvious signs of oxidation are changes in the nutritional quality of food products. For example, when oxidation occurs, fat-soluble vitamins such as A, D, and E are destroyed: the caloric content of foods can also decrease and essential fatty acids such as linoleic acid are lost.

1.3 Food-grade antioxidants

1.3.1 *Mechanism and functions*

Antioxidants for fats and oils function by interfering in the formation of the free radicals that initiate and propagate oxidation. The following chemical reactions describe the antioxidant mechanism.

Reaction (5) delays or inhibits the initiation step of free radical oxidation. Reaction (6) inhibits the propagation step.

Antioxidants must form 'stable', low-energy free radicals that will not further propagate the oxidation of fats and oils. All the antioxidants under discussion in this chapter are phenol derivatives (7).

$$\text{AH} + \left[\cdots C-\overset{\overset{\textstyle H}{|}}{\underset{\textstyle \bullet}{C}}-C=C-C \cdots\right] \rightarrow A\bullet + \left[\cdots C-\overset{\overset{\textstyle H}{|}}{\underset{\textstyle H}{C}}-C=C-C \cdots\right]$$

Antioxidant   Fat Free Radical   Antioxidant Free Radical   Original Fat Molecule

(5)

$$\text{AH} + \left[\cdots C-\overset{\overset{\textstyle H}{|}}{\underset{\textstyle \underset{\textstyle \bullet}{\overset{\textstyle |}{O}}}{\underset{\textstyle |}{O}}}-C=C-C \cdots\right] \rightarrow A\bullet + \left[\cdots C-\overset{\overset{\textstyle H}{|}}{\underset{\textstyle \underset{\textstyle H}{\overset{\textstyle |}{O}}}{\underset{\textstyle |}{O}}}-C=C-C \cdots\right]$$

Antioxidant   Peroxide Free Radical   Antioxidant Free Radical   Hydroperoxide

(6)

Phenol   BHA   BHT   TBHQ

Propyl Gallate   Gamma Tocopherol

(7)

The underlying phenolic structure of the major food-grade antioxidants allows them to form low-energy free radicals through resonance hybrids. This is chemically depicted in (8).

A knowledge of the mechanism of antioxidant performance reinforces several important aspects of antioxidant usage.

1. Antioxidants are not oxygen scavengers or absorbers. They function by reaction with free radicals.

Phenol   Free Radical   Stable Resonance Hybrid

(8)

2. Antioxidants must be added to fats, oils and food products as early as possible for maximum benefit. Addition of antioxidant to fats and oils with a substantial peroxide content will result in loss of antioxidant performance. This effect is demonstrated in tests with TBHQ in peanut oil (Table 1.2).
3. Antioxidants cannot rejuvenate an oxidised fat or oil.

Table 1.2 Antioxidant performance – fresh vs. oxidised oil

Antioxidant treatment		Initial peroxide value[1]	AOM stability[2]
Peanut oil + TBHQ	0.02%	0	102
Aged peanut oil[3] + TBHQ	0.02%	3	76

[1]Milliequivalents of peroxide per kilogram of oil.
[2]Hours to develop peroxide value of 70 meq kg^{-1} of oil.
[3]Same original peanut oil sample.

1.3.2 Individual properties of major antioxidants

All antioxidants have strengths and weaknesses and present problems as well as provide solutions. The individual properties of the major antioxidants deserve closer attention.

BHA (butylated hydroxyanisole). Chemically, BHA is a mixture of two isomers, as shown in (9).

3-tertiarybutyl-4-   2-tertiarybutyl-4-
hydroxyanisole   hydroxyanisole

(9)

The 3-isomer is the better antioxidant. Commercial BHA generally has at least 90% 3-isomer content.

Facts about BHA (Table 1.3)

1. BHA is a white waxy solid.
2. BHA is sold in flake or tablet form.
3. BHA is effective in animal fats and relatively ineffective in unsaturated vegetable oils.
4. BHA provides good 'carry through' potency, especially in animal fats in baked foods. 'Carry through' refers to an antioxidant's ability to be added to a food component, survive a processing step such as frying or baking, and imparting stability to the finished food product.
5. BHA is highly fat soluble and insoluble in water.
6. BHA is volatile and steam distillable. Its volatility makes BHA useful for addition to packaging materials.

Table 1.3 Physical properties of BHA

Molecular weight	180.25
Physical appearance	White waxy tablets or flakes
Boiling range, 733 mm Hg, °C	264–270
Melting range, °C	48–63
Odour	Slight
Solubility	

Solvent	Temperature (°C)	Solubility (%, approx.)
Water	0–50	Insoluble
Glycerol	25	1
Ethyl alcohol	25	50+
Diisobutyl adipate	25	60
Propylene glycol	25	70
Glyceryl monooleate	25	50
Coconut oil	25	40
Cottonseed oil	25	30
Corn oil	25	40
Peanut oil	25	40
Soybean oil	25	50
Lard	50	50
Yellow grease	50	50
Paraffin	60	60+
Mineral oil	25	5

(10)

BHT (*butylated hydroxytoluene*). Chemically, BHT (**10**) is 2,6-ditertiary-butyl-4-methyl phenol.

Facts about BHT (Table 1.4)

1. BHT is a white crystalline solid.
2. BHT has properties similar to BHA.
3. BHT provides good synergism when used in combination with BHA.
4. BHT is widely used as an industrial antioxidant and is, therefore, relatively inexpensive.

Table 1.4 Physical properties of BHT

Molecular weight	220
Physical appearance	White granular crystal
Boiling point, 760 mm Hg, °C	265
Melting point, °C	69.7
Odour	Very slight
Solubility	

Solvent	Temperature (°C)	Solubility (%, approx.)
Water	0–60	Insoluble
Glycerol	25	Insoluble
Ethyl alcohol	25	25
Diisobutyl adipate	25	40
Propylene glycol	0–25	Insoluble
Glyceryl monooleate	25	15
Coconut oil	25	30
Cottonseed oil	25	30
Corn oil	25	30
Olive oil	25	25
Peanut oil	25	30
Soya oil	25	30
Lard	50	40
Yellow grease	50	40
Paraffin	60	60+
Mineral oil	25	5

Propyl gallate. Chemically, propyl gallate (**11**) is the propanol ester of 3,4,5-tri-hydroxybenzoic acid.

Facts about propyl gallate (Table 1.5)

1. Propyl gallate is a white crystalline solid.

(**11**)

Table 1.5 Physical properties of propyl gallate

Molecular weight	212.20
Physical appearance	White crystalline powder
Boiling point, 760 mm Hg, °C	Decomposes above 148
Melting range, °C	146–150
Odour	Very slight
Solubility	

Solvent	Temperature (°C)	Solubility (%, approx.)
Water	25	<1
Glycerol	25	25
Ethyl alcohol	25	60+
Diisobutyl adipate	25	15
Propylene glycol	25	55
Glyceryl monooleate	25	5
Cottonseed oil	25	1
Corn oil	25	Insoluble
Peanut oil	25	<1
Soybean oil	85	2
Lard	50	1
Yellow grease	50	1
Mineral oil	25	<1

2. Propyl gallate imparts good stability to vegetable oils.
3. Propyl gallate provides good synergism with BHA and BHT.
4. Propyl gallate is heat sensitive and decomposes at its melting point of 148°C.
5. Propyl gallate has poor 'carry through' properties.
6. Propyl gallate has poor fat solubility and some water solubility.
7. Propyl gallate can form coloured complexes with metal ions that can adversely affect the appearance of fats and oils.

TBHQ (tertiary butylhydroquinone). Chemically, TBHQ (**12**) is tertiary butylhydroquinone.

Facts about TBHQ (Table 1.6)

1. TBHQ is a white to light tan powder.
2. TBHQ is the most effective antioxidant for most fats and oils, especially vegetable oils.
3. TBHQ provides poor 'carry through' in baking applications. Combinations of TBHQ and BHA give improved performance.

(12)

Table 1.6 Physical properties of TBHQ

Molecular weight	166.22
Physical appearance	White to light tan crystals
Boiling point, 760 mm Hg, °C	300
Melting range, °C	126.5–128.5
Odour	Very slight
Solubility	

Solvent	Temperature (°C)	Solubility (%, approx.)
Water	25	<1
Water	95	5
Ethanol	25	60
Ethyl acetate	25	60
Propylene glycol	25	30
Glyceryl monooleate	25	10
Corn oil	25	10
Cottonseed oil	25	10
Safflower oil	25	5
Soybean oil	25	10
Lard	50	5

4. TBHQ provides excellent 'carry through' in frying applications.
5. TBHQ has adequate fat and oil solubility.
6. TBHQ does not form coloured complexes with iron and copper, but can turn pink in the presence of bases.

Tocopherols. The chemical structures of the tocopherols used as antioxidants are shown in structure (**13**) and their physical properties are given in Table 1.7.

alpha-tocopherol

gamma-tocopherol

delta-tocopherol

(**13**)

Table 1.7   Physical properties of tocopherols

Physical form	Slightly viscous liquid
Colour	Red to reddish brown
Odour	Slight, mild
Taste	Slight, mild
Solubility	
Vegetable oils	Soluble
Fats	Soluble
Essential oils	Soluble
Ethanol	Soluble
Water	Insoluble

Tocopherols enjoy the greatest popularity of the natural source antioxidants. They occur naturally in vegetable oils, the prime commercial source being soybeans. They are extracted from tocopherol-rich deodoriser distillates which are by-products from steam stripping of vegetable oils during their final processing step.

α-Tocopherol is primarily recognised as a source of Vitamin E and is available as a natural product (d-α-tocopherol) or a synthetic product (dl-α-tocopherol). α-Tocopherols exhibit some antioxidant potency, but the γ- and δ-tocopherol epimers are considered to be substantially more effective antioxidants (Table 1.8).

Table 1.8   Antioxidant performance of natural mixed tocopherols vs. synthetic dl-α-tocopherol in lard

Antioxidant treatment	Total tocopherols (%)	AOM stability[1]
Control		7
dl-α-tocopherol	0.02	33
Mixed tocopherols	0.005	33
Mixed tocopherols	0.02	48

[1]Hours to develop peroxide value of 20 meq kg^{-1} of oil.

The products that are most appropriate for use as antioxidants usually contain γ- and δ-tocopherols at a minimum of 80% of the total tocopherol concentration. These mixed tocopherol concentrates are valuable ingredients when regulations do not permit the use of more effective synthetic antioxidants or where natural source antioxidants are simply preferred.

Most fats, oils, and food products derived from animal origin are low or deficient in natural antioxidants such as tocopherols, lecithin, citric acid, and ascorbic acid. Examples of such food products are lard, tallow, poultry fat, essential oils and paraffin wax. Tocopherols are most applicable to these products. Oils and food products from vegetable origins generally contain sufficient tocopherols for good stability. Unless the natural antioxidants in these products have been lost during refining and processing or by thermal degradation, the benefits of adding tocopherols to such products are marginal.

1.4 Acid synergists

Acid synergists are often added to fats and oils in combination with primary antioxidants to improve their antioxidant effectiveness. The more important acid synergists are citric acid and its ester, monoglyceride citrate; and ascorbic acid and its ester, ascorbyl palmitate. The lipophilic esters of citric acid and ascorbic acid have improved fat and oil solubility.

In general, citric acid and its esters are used in combination with the synthetic antioxidants and ascorbic and its esters are used with tocopherols.

The benefits of acid synergists are easily demonstrated. Less obvious, however, is the mechanism by which the synergists perform. Synergists may enhance antioxidant effectiveness by one or more of the following mechanisms.

1. Provide an acidic medium to improve stability of primary antioxidants and fats and oils.
2. Regenerate primary antioxidants.
3. Chelate and deactivate pro-oxidant metal contaminants such as iron and copper.
4. Scavenge oxygen (ascorbic acid).

1.5 Antioxidant formulations (solutions)

Antioxidant solutions are widely used in the food industry. These formulations offer both performance advantages and 'ease of use' features. Most formulations usually contain one or more of the major antioxidants, along with an acid synergist dissolved in a food-grade solvent. These solvents include vegetable oils, propylene glycol, glyceryl monooleate, ethanol and acetylated monoglycerides. Combinations of solvents are also used to improve solubility and dispersibility of the antioxidants.

Antioxidant formulations perform the following functions that benefit the prospective user of antioxidants.

1. Combine performance characteristics of individual antioxidants.
2. Provide ease of use and handling.
3. Improve accuracy of application.
4. Allow combination of antioxidants and acid synergists in one product.
5. Enhance the solubility and dispersibility of antioxidants.

6. Minimise discoloration tendencies of antioxidants.
7. Provide synergistic combinations of antioxidants.

1.6 Selection of antioxidants

Many considerations are required for selection of the best antioxidant system for a specific application. The more important ones are:

1. Type of potency desired.
   (a) Animal fats
   (b) Vegetable oils
2. Is 'carry through' effectiveness required?
3. Solubility or dispersibility of the antioxidant.
4. Discoloration tendencies.
5. pH of the food product.
6. Type of food processing: frying, baking, spray-drying, extrusion, etc.
7. Flavour and odour.
8. Legal and regulatory status.

1.7 Methods of antioxidant addition

Antioxidants are generally incorporated into foods by direct addition to fats and oils. This is probably the most convenient and effective method of addition. Antioxidants work best when added as early as possible. Fat and oil suppliers are generally adept at adding antioxidants. It may be easier for the supplier to add antioxidant than the food manufacturers.

Spray addition is also a convenient way to add antioxidants. Antioxidants can be dissolved in a food-grade solvent such as propylene glycol, ethanol or vegetable oil and sprayed on products such as nuts and cereals.

Andioxidants can be added to foods via the packaging. Antioxidants can be incorporated into packaging materials such as wax paper, polyethylene, and paperboard. The antioxidant, through volatilisation, can penetrate and provide protection to food products contained within the package. In addition, most packaging material is subject to oxidation and benefits from the antioxidant's presence.

Proper incorporation of antioxidants requires the following considerations:

1. Make sure the antioxidant is thoroughly dissolved and homogeneously dispersed in the finished product. Failure to do this is the most prevalent mistake in antioxidant usage.
2. Add the antioxidant at the proper time.
3. Be aware that processing steps can remove or destroy antioxidants.
4. Add the legal amount of antioxidant.

1.8 Evaluation of antioxidant effectiveness

Oxidative stability of antioxidant-treated fats and oils can be determined by storing the product under normal use conditions and examining the samples periodically for changes in odour or flavour, or by testing them chemically for rancidity. These procedures generally take too long and accelerated tests are more often used to compare untreated (controls) and antioxidant-treated products. The following two test methods are particularly useful for quality control or product development purposes. These test methods are the basis of the performance data shown in the following section on major applications of food antioxidants.

1.8.1 *Active oxygen method* (AOM)

The AOM test is widely used on fats and oils which are liquid at the test temperature. It is not applicable to solid materials. In the AOM test, air is bubbled through the heated test sample to accelerate oxidation and shorten the test time. Periodic analyses are conducted to determine when the peroxide content reaches the rancidity point which is generally 20 meq kg^{-1} of oil in animal fats and 70 meq kg^{-1} of oil in vegetable oils.

1.8.2 *Oven storage tests* (*Schaal oven test*)

Higher temperature will accelerate the oxidation process. Oven storage tests are simply shelf storage tests conducted at elevated temperatures to speed up the procedure. A common test temperature is 62.8°C (145°F). Periodic odour and flavour evaluations are commonly used, but chemical analyses (for example, peroxide content) may be used to determine rancidity development in the samples.

1.9 Performance of antioxidants in major food applications

The data in this section will demonstrate the increased stability that can be achieved by adding food antioxidants to fats, oils and food products. In addition, comparative performance of the individual antioxidants will be shown in various applications.

1.9.1 *Vegetable oils*

Almost all vegetable oils respond well to antioxidant treatment. A compilation of all the performance data generated during the last two decades would indicate the following relative effectiveness of synthetic antioxidants in vegetable oils:

$$TBHQ > propyl\ gallate > BHT > BHA$$

Tocopherols are generally not used in vegetable oils when food regulations permit the application of the more effective synthetic antioxidants.

Antioxidant performance data for the world's major vegetable oils are shown in Figures 1.1 to 1.17 and in Table 1.9.

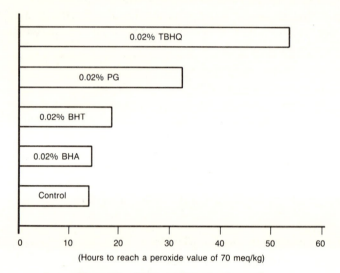

(Hours to reach a peroxide value of 70 meq/kg)

Figure 1.1 AOM stability of soybean oil.

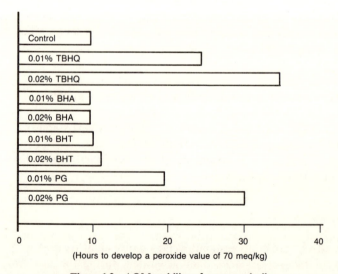

(Hours to develop a peroxide value of 70 meq/kg)

Figure 1.2 AOM stability of cottonseed oil.

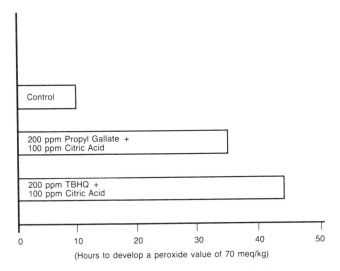

Figure 1.3 AOM stability of cottonseed oil.

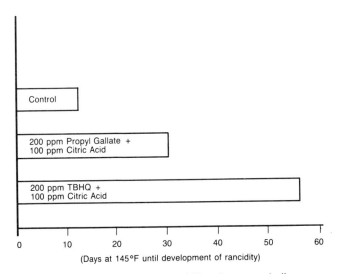

Figure 1.4 Oven storage stability of cottonseed oil.

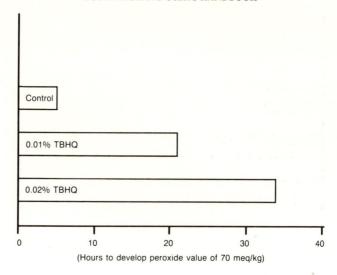

Figure 1.5　AOM stability of sunflower seed oil.

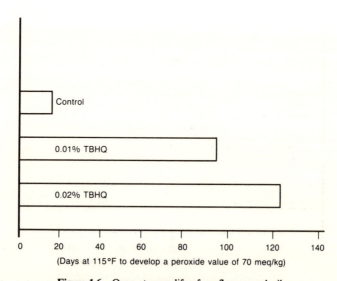

Figure 1.6　Oven storage life of sunflower seed oil.

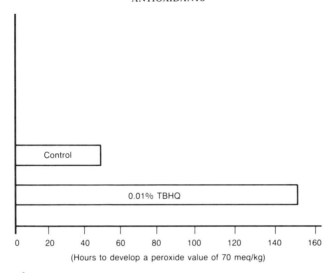

Figure 1.7 AOM stability of palm oil.

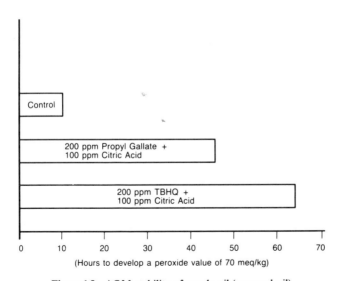

Figure 1.8 AOM stability of canola oil (rapeseed oil).

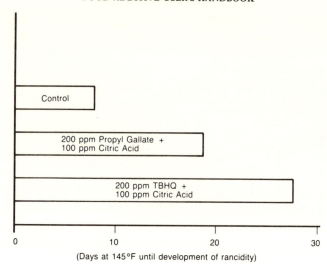

Figure 1.9   Oven storage stability of canola oil (rapeseed oil).

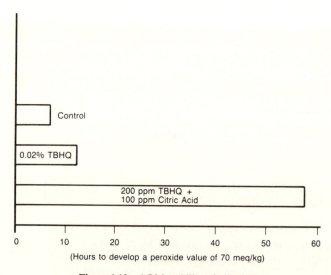

Figure 1.10   AOM stability of olive oil.

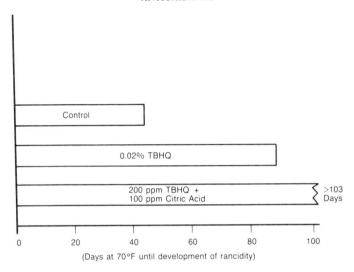

Figure 1.11   Shelf storage life of olive oil.

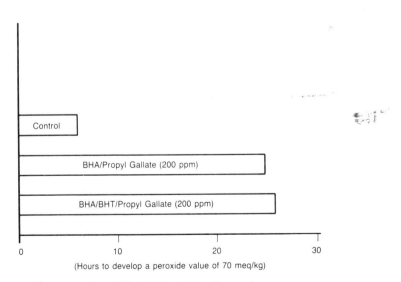

Figure 1.12   AOM stability of corn oil.

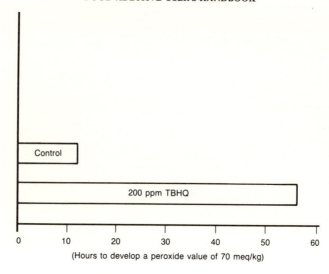

Figure 1.13   AOM stability of corn oil.

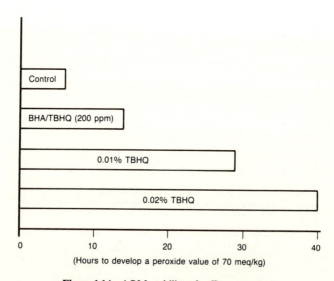

Figure 1.14   AOM stability of safflower seed oil.

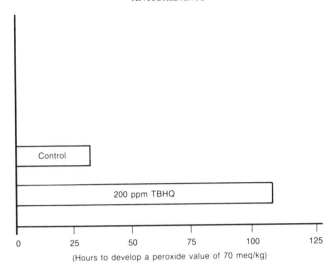

Figure 1.15   AOM stability of high-oleic safflower seed oil.

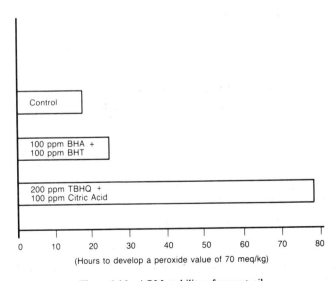

Figure 1.16   AOM stability of peanut oil.

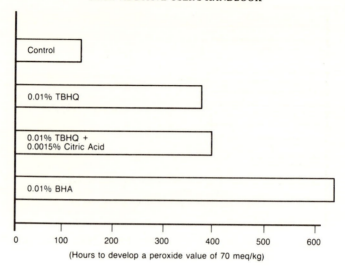

(Hours to develop a peroxide value of 70 meq/kg)

Figure 1.17 AOM stability of coconut oil.

Table 1.9 Antioxidant performance in canola oil

Antioxidant treatment		AOM stability[1]
Control		8
TBHQ	0.02%	62
Propyl gallate	0.02%	35
BHA/Propyl gallate combination	0.02%	22
BHA/BHT/Propyl gallate combination	0.02%	22

[1]Hours to develop peroxide value of 70 meq kg^{-1} of oil.

1.9.2 Animal fats and oils

Animal fats are deficient in natural antioxidants and have a low inherent oxidative stability. The addition of antioxidants is often essential for adequate shelf-life for animal fats and finished food products containing them.

Because animal fats contain little or no natural antioxidants, the addition of tocopherols gives a sizeable increase in oxidative stability. Acid synergists are also important in stabilising animal fats.

The general performance trend of antioxidants in animal fats and oils is:

$$TBHQ > propyl\ gallate > BHA > BHT > tocopherols$$

The performance data given in Tables 1.10–1.13 and Figures 1.18–1.20 demonstrate the capability of antioxidants to increase the shelf-life of animal fats and oil.

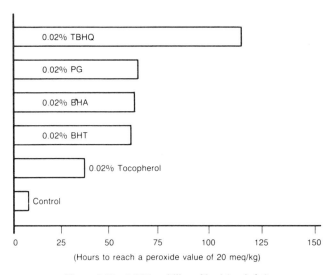

Figure 1.18 AOM stability of lard (pork fat).

Table 1.10 Effect of tocopherols and acid synergists on storage stability of lard

Antioxidant treatment		Storage stability[1]
Control		45
Tocopherols	0.01%	210
Tocopherols +citric acid	0.01% 0.005%	294

[1]Days at room temperature to develop peroxide value of 20 meq kg^{-1} of oil.

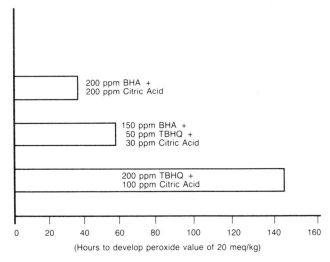

Figure 1.19 AOM stability of tallow (beef).

Table 1.11   Performance of mixed tocopherols in beef tallow

Antioxidant treatment		AOM stability[1]	Oven stability[2]
Control		7	15
Tocopherols	0.02%	105	76

[1]Hours to develop peroxide value of 20 meq kg^{-1} of oil.
[2]Days to rancidity at 62.8°C (145°F).

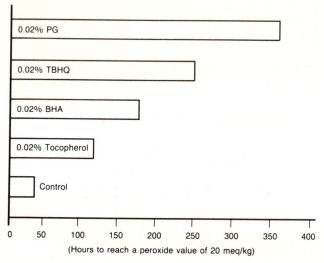

Figure 1.20   AOM stability of butterfat (anhydrous).

Table 1.12   Antioxidant performance in poultry fat

Antioxidant treatment		AOM stability[1]
Control		5
TBHQ	0.005%	27
TBHQ	0.010%	39
TBHQ	0.020%	56
BHA	0.005%	12
BHA	0.010%	15
BHA	0.020%	18
BHT	0.005%	10
BHT	0.010%	14
BHT	0.020%	20
Propyl gallate	0.005%	13
Propyl gallate	0.010%	19
Propyl gallate	0.020%	30

[1]Hours to develop peroxide value of 20 meq kg^{-1} of oil.

Table 1.13 Antioxidant performance[1] in menhaden oil

Antioxidant treatment		Oxidative stability[2] (minutes)
Control		37
Tocopherols	0.02%	63
TBHQ	0.02%	124
TBHQ	0.02%	265
+ Tocopherols	0.02%	

[1]Determined by differential scanning calorimetry.
[2]Stability is time to oxidise at 80°C in an O_2 atmosphere.

1.9.3 Essential oils and flavourings

Flavourings suffer from oxidative changes with the loss of their intense flavour 'high notes'. All of the major antioxidants offer a measure of protection to most flavours and essential oils (Tables 1.14 and 1.15).

Table 1.14 Antioxidant performance in flavourings

Antioxidant treatment		Storage stability[1]					
		Lemon		Orange		Peppermint	
		29.4°C (85°F)	37.8°C (100°F)	29.4°C (85°F)	37.8°C (100°F)	29.4°C (85°F)	37.8°C (100°F)
Control		59	60	40	39	94	20
TBHQ	0.005%	126	127	82	73	122	25
TBHQ	0.010%	145[2]	125	75	103	124	133
TBHQ	0.020%	—[3]	—	91	65	107	72
BHA	0.005%	120	87	118	93	68	42
BHA	0.010%	133	96	217	103	42	—
BHA	0.020%	—	—	366	142	64	33
BHT	0.005%	62	103	122	72	100	74
BHT	0.010%	75	111	228	103	85	100
BHT	0.020%	—	—	451	159	104	114
Propyl gallate	0.005%	72	116	104	93	138	39
Propyl gallate	0.010%	64	118	102	103	86	129
Propyl gallate	0.020%	—	—	95	105	121	125

[1]Days to develop rancidity.
[2]Samples not rancid at end of test period.
[3]Dash indicates that tests were not conducted.

Table 1.15 Effectiveness of tocopherols in citrus oils

| Tocopherols | AOM stability[1] | | | |
	Lemon terpenes	Grapefruit oil	Orange oil	Lemon oil
Control	3	35	32	18
0.01%	98	60	75	57
0.025%	159	96	119	132
0.04%	127	146	174	188

[1]Hours at 65°C to develop peroxide value of 70 meq kg^{-1} of oil.

1.9.4 *Frying applications*

Fried foods, such as potato chips and corn chips, may contain as much as

Table 1.16 Stability of potato chips fried in TBHQ-treated oils

Frying oil	Antioxidant treatment		Storage stability[1]
Cottonseed	Control		11
	TBHQ	0.02%	23
Soybean	Control		8
	TBHQ	0.02%	25
Safflower	Control		4
	TBHQ	0.02%	23
Sunflower	Control		13
	TBHQ	0.02%	35

[1]Days at 62.8°C (145°F) to develop rancidity.

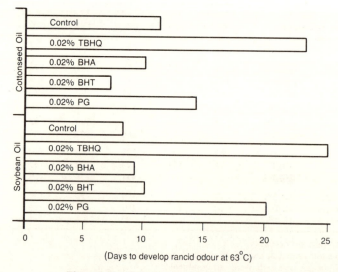

(Days to develop rancid odour at 63°C)

Figure 1.21 Oven storage life of potato chips.

50% oil. These products also have a large surface area. Rancidity development is always a threat in such products.

TBHQ (Table 1.16) is generally the best antioxidant for protecting frying
oils against oxidation, and it provides carry-through protection to the
finished fried product.

Where TBHQ is not permitted, various combinations of BHA, BHT and
propyl gallate provide synergism (Figures 1.21–1.23).

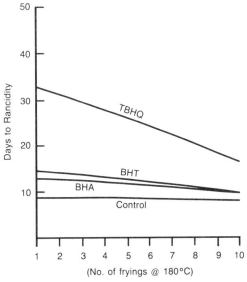

Figure 1.22   Effect of multiple fryings on the oven stability of potato chips fried in soybean oil.

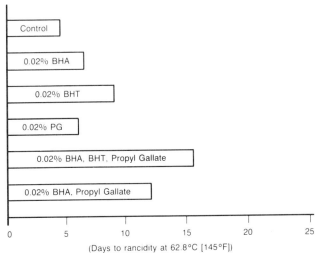

Figure 1.23   Shelf-life of potato chips prepared in safflower oil.

Tocopherols are useful in frying applications that use animal fat and oils because of their heat stability and their resistance to volatilisation. Tocopherols provide excellent 'carry through' effectiveness to fried foods (Table 1.17).

Table 1.17 Stability of potato chips fried in beef tallow

Antioxidant treatment	Storage stability[1]
Control	6
Tocopherols 0.02%	65

[1] Days at 62.8°C (145°F) to develop rancidity.

1.9.5 Baking applications

BHA, BHT and tocopherols are added to fat and oils used in baking applications because their 'carry through' effectiveness imparts stability to the finished baked good (Tables 1.18–1.20). However, it is often desirable to have TBHQ or propyl gallate in the fat or oil to provide storage stability prior to its usage in baking. For this reason, synergistic combinations of BHA, BHT and propyl gallate, or BHA, BHT and TBHQ, are often used to stabilise fats and oils used for baking.

Table 1.18 Performance of antioxidants in bakery products

Product	Shortening	Antioxidant treatment		Storage stability
Cookies	Soybean oil	Control		7
		BHA	0.0098%	21
		+ Propyl gallate	0.0042%	
Pastry mix	Lard	Control		30
		BHA	0.02%	72
Bread	Lard	Control		10
		BHA	0.01%	20
		+ Propyl gallate	0.003%	
Pastry	Lard	Control		2
		BHA	0.02%	27
		BHA	0.01%	38
		+ TBHQ	0.01%	
Crackers	Lard	Control		3
		BHA	0.02%	33

[1] Days at 62.8°C (145°F) to develop rancidity.

Table 1.19 Carry-through effectiveness of mixed tocopherols in bakery products

Product	Shortening	Antioxidant treatment		Storage stability[1]
Pastry	Beef tallow	Control		28
		Tocopherols	0.02%	61
Crackers	Beef tallow	Control		14
		Tocopherols	0.02%	61

[1] Days at 62.8°C (145°F) to develop rancidity.

Table 1.20 Carry-through effectiveness of antioxidants in bakery products containing lard

		Storage stability[1]			
		Pastry		Crackers	
Antioxidant treatment		37.8°C (100°F)	62.8°C (145°F)	37.8°C (100°F)	62.8°C (145°F)
Control		10	2	9	3
TBHQ	0.005%	14	2	27	7
TBHQ	0.010%	25	3	—	10
BHA	0.005%	35	8	125	12
BHA	0.010%	96	21	218	22
BHT	0.005%	40	5	90	10
BHT	0.010%	—	10	132	14
Propyl gallate	0.005%	23	2	—	3
Propyl gallate	0.010%	28	5	33	6
TBHQ +BHA	0.010% 0.010%	236	38	236	23
TBHQ +BHT	0.010% 0.010%	167	16	255	25

[1] Days to develop rancidity.

1.9.6 Nut products

Nut products are widely used alone as snack foods and as important components of other products such as confections and cereals. All the synthetic antioxidants provide good protection to nut kernels and pieces (Figure 1.24 and Tables 1.21 and 1.22). TBHQ is generally more effective in nut oils and nut butters.

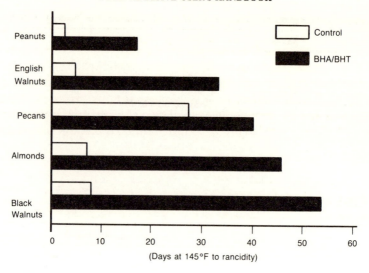

Figure 1.24   Effect of BHA/BHT treatment on stability of nut kernels.

Table 1.21   Performance of antioxidants in nuts

Nut	Method of antioxidant addition	Antioxidant treatment		Storage stability[1]
Pecans	Added to roasting oil	Control		24
		TBHQ	0.02%	78
Peanuts	Added to roasting oil	Control		9
		TBHQ	0.02%	35
Peanuts	Added to salt	Control		6
		BHA	0.005%	23
		+BHT	0.005%	
English walnuts	Spray addition	Control		5
		BHA	0.01%	33
		+BHT	0.01%	

[1] Days at 62.8°C (145°F) to develop rancidity.

Table 1.22   Performance of antioxidants in peanut paste (butter)

Antioxidant treatment[1]		Storage stability[2]
Control		28
BHA	0.02%	28
BHT	0.02%	57
Propyl gallate	0.02%	40
TBHQ	0.02%	125

[1] Based on oil content.
[2] Days at 62.8°C (145°F) to develop peroxide value of 70 meq kg^{-1} of oil.

1.9.7 *Grains and cereals*

Cereals are widely used as ingredients in foods such as candies and confections, snack foods (granola) and dry breakfast cereals. Although cereals are low in lipid content, the lipids are very unsaturated and quite unstable. Additional thermal processing steps such as extrusion, roasting, toasting and baking decrease the stability of cereals and grains. All the major antioxidants provide protection to grains and cereals (Table 1.23).

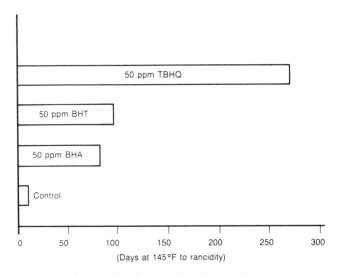

Figure 1.25   Effectiveness of antioxidants in mixed cereal (corn, wheat, and rice).

Table 1.23   Performance of tocopherols in cereals and grains

Product	Tocopherol treatment		Storage stability[1]
Oat flour	Control		48
	Tocopherols	0.02%	96
Oat and	Control		78
wheat flour	Tocopherols	0.02%	100
Granola	Control		18
	Tocopherols	0.02%	>33

[1] Days at 62.8°C (145°F) to develop rancidity.

Dry breakfast cereals are often stabilised by inclusion of antioxidants (especially BHA and BHT) into the packaging materials (Figure 1.25). BHA and/or BHT, having appreciable vapour pressure, will volatilise from the packaging and permeate the cereal thus providing protection.

1.9.8 *Crude vegetable oils*

Although most crude vegetable oils benefit from the protection provided by their natural antioxidant content of tocopherols, phospholipids and carotene, they can still suffer serious deterioration during storage, transit and refining. The protective benefits of adding TBHQ to crude oil have been demonstrated in major oils including soy oil (Figure 1.26), palm oil (Figures 1.27 and 1.28), and sunflower oil (Figure 1.29). These benefits include:

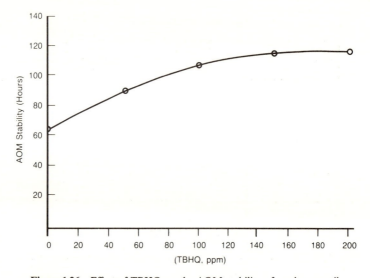

Figure 1.26   Effect of TBHQ on the AOM stability of crude soya oil.

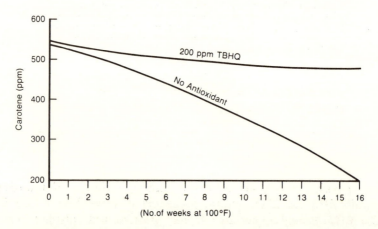

Figure 1.27   Effect of TBHQ on the carotene content of crude palm oil.

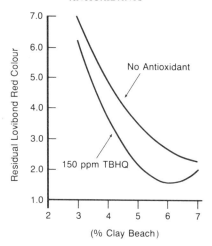

Figure 1.28 Effect of TBHQ on bleachability of palm oil.

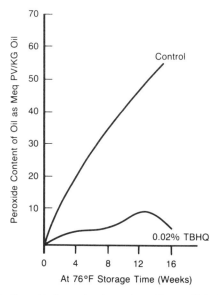

Figure 1.29 Peroxide formation in crude sunflower oil during storage.

1. Inhibition of peroxide formation.
2. Protection of carotene.
3. Protection of tocopherols.

These antioxidant benefits translate into economic benefits. These include:

1. Increased storage times.
2. Lower refining losses.

3. Reduced usage of bleaching clay.
4. Improved stability of refined oil.

When crude oils treated with TBHQ are subjected to the refining step of steam deodorisation, the TBHQ is completely removed and the resulting refined oil is antioxidant free.

1.9.9 *Food packaging*

Modern-day packaging must not only be aesthetically pleasing, but also functional in prolonging shelf-life. In addition to providing a sanitary enclosure, packaging can serve as oxygen and moisture barriers and inhibit transmission of light. Packages can also serve as a convenient way of adding antioxidants to foods.

Table 1.24   Effect of antioxidant treated-packaging on the stability of peanut butter candy

Antioxidant treatment	Storage stability[1]
Control	70
BHA[2] 0.05%	220

[1] Days at room temperature to develop rancidity.
[2] Added to packaging material.

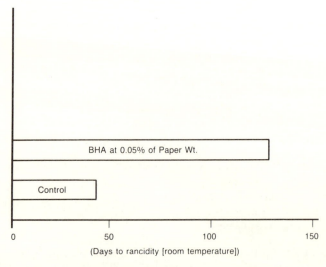

Figure 1.30   Effectiveness of antioxidant in packaging material (chocolate-almond candy in glassine paper).

Antioxidants can be added to waxes, paraffin and polymers used in food packaging and allowed to volatilise into the enclosed food products. Although this is not the most efficient way to add antioxidants, it is very convenient. BHA and BHT are effective in packaging applications because of their volatility.

Some packaging materials are also subject to oxidation. Antioxidants also help provide for their protection (Table 1.24). Antioxidants are applied to a wide variety of packaging materials, including glassine paper (Figure 1.30), paperboard, cartonboard, chocolate board, and plastic film.

1.9.10 *Confections*

Many of the ingredients that are used in confections are subject to deterioration. Dairy products, fats and oils, nutmeats and essential oils all tend to undergo various types of degradation, which leads to a loss of fresh taste and flavour or the development of off-flavours. Poor stability in any of the ingredients can cause spoilage of the final product.

Antioxidants are widely used in confections to increase shelf-life. The addition of antioxidants to many confectionery ingredients has been discussed in previous sections of this chapter. The prospective user of antioxidants can refer to this information in selecting the right antioxidant for a specific confection.

1.9.11 *Meat products*

A principal route of deterioration in the quality of meat and meat products (both raw and cooked) is through oxidation of the lipid or fatty portions. Various tests have shown that food antioxidants can inhibit oxidative deterioration in beef, poultry, pork and fish.

One of the major problems with treating meat products is achieving intimate contact of the antioxidant with meat lipids. For this reason, antioxidants perform more efficiently when added to diced, minced, flaked or ground products.

Some processed meat products, such as pork sausage, are found to be especially prone to oxidative deterioration even under freezer storage conditions where oxidative rancidity development is strongly inhibited in most food products. This strong tendency towards oxidation is due to several factors:

1. Grinding of the meat results in greatly increased exposure of the lipids to oxidation.
2. Haem pigments in the meat are brought into contact with and serve to catalyse oxidation of the lipids.

3. High levels of salt strongly catalyse the oxidation reactions, especially under freezer storage conditions.

BHA, combined with citric acid as a metal chelating agent, is effective in improving oxidative stability of fresh pork sausage, as demonstrated by oxygen bomb evaluations (Table 1.25).

Table 1.25 Performance of BHA in fresh pork sausage

Antioxidant treatment		Oxygen bomb stability[1]
Control		< 5
BHA	0.01%	11
+Citric acid	0.01%	

[1] Hours at 100°C and 100 lb in^{-2} oxygen pressure.

Lipid oxidation is also found to be a major cause of rapid deterioration of freeze-dried meats. Studies have indicated that antioxidants are effective in improving the keeping quality of freeze-dried ham (Table 1.26).

Table 1.26 Performance of BHA in freeze-dried ham

Antioxidant treatment		Storage stability[1]	
		37.8°C (100°F)	62.8°C (145°F)
Control		3.5	2
BHA[2]	0.02%	33.5	8.5

[1] Days to develop rancidity.
[2] Antioxidant solution sprayed on dried ham.

1.9.12 Feed fats

Inedible tallows and greases are important by-products of the meat-packing industry and other food-related industries. They find widespread use in the formulation of high-energy animal and poultry feeds. These fats provide more than twice as many calories per pound as protein or carbohydrate and are important ingredients in many types of animal feeds such as dry or moist pet foods, poultry feeds, calf starters, milk replacers, swine feeds, and cattle rations.

Three factors that affect the choice of antioxidants for inedible fats are the quality of the raw materials from which the fat is derived, the processing conditions, and the types and concentrations of trace metals in the fat. With these factors in mind, the applicability of each antioxidant can be assessed based on its particular properties and performance (Table 1.27).

Effectiveness of antioxidants in stabilising various grades of greases and tallows has been determined by laboratory tests. Industry-wide standards of stability for tallows and greases intended for use in animal feeds have not been officially established; however, many feed manufacturers specify a minimum fat stability of 20 AOM hours at the time of addition to the feed (Table 1.28).

Table 1.27 Performance of antioxidants in inedible fats

Fat	Antioxidant treatment		AOM stability[1]
Grease	Control		12
	BHA	0.01%	51
Inedible fat	Control		10
	BHA	0.01%	25
Grease	Control		6
	BHA	0.01%	44
	+ Citric acid	0.01%	
Yellow grease	Control		3
	BHA	0.02%	37
	+ Citric acid	0.02%	
White grease	Control		2
	BHA	0.02%	41
	+ Citric acid	0.02%	
Feed-grade tallow	Control		2
	BHA	0.02%	137
	+ Citric acid	0.02%	

[1] Hours to develop peroxide content of 20 meq kg^{-1} of oil.

Table 1.28 Performance of food-grade antioxidants vs. ethoxyquin[1] in feed fats

Fat	Antioxidant treatment		AOM stability[2]
Poultry	Control		3
Poultry	Ethoxyquin	0.088%	16
Poultry	BHA	0.01%	15
	+ Citric acid	0.01%	
Poultry	TBHQ	0.01%	60
	+ Citric acid	0.005%	
Tallow	Control		5
Tallow	Ethoxyquin	0.088%	24
Tallow	BHA	0.01%	77
	+ Citric acid	0.01%	
Tallow	TBHQ	0.01%	85
	+ Citric acid	0.005%	

[1] Widely used antioxidant approved for feed fats only.
[2] Hours to develop peroxide value of 20 meq kg^{-1} of fat.

Table 1.29 Food and Drug Administration (FDA) regulations (CFR = Code of Federal Regulations – USA)

Application	Food additive regulation	Permitted antioxidant	Limitation or tolerance (percent by weight of food unless otherwise specified)
General uses	21 CFR 182.3169	BHA	0.02% (200 ppm) singly or in combination, by weight, in fat or oil portion of food including the essential (volatile) oil except where prohibited by Standard of Identity. TBHQ is lawful for use in combination only with BHA and/or BHT
	21 CFR 182.3173	BHT	
	21 CFR 184.1660	Propyl gallate	
	21 CFR 172.185	TBHQ	
	21 CFR 182.3890	Tocopherols	Good manufacturing practice
SPECIFIC FOODS			
Chewing gum base	21 CFR 172.615	BHA, BHT, propyl gallate	0.1% (1000 ppm) singly or in combination
Synthetic flavouring substances and adjuvants	21 CFR 172.515	BHA	0.5% (5000 ppm) of flavouring oil
Active dry yeast	21 CFR 172.110	BHA	0.1% (1000 ppm)
Beverages and desserts from dry mixes			0.0002% (2 ppm)
Dry mixes for beverages and desserts			0.009% (90 ppm)
Dry diced glazed fruit			0.0032% (32 ppm)
Dry breakfast cereals	21 CFR 172.110 and/or	BHA and/or BHT	0.005% (50 ppm)
Emulsion stabilisers for shortening	21 CFR 172.115		0.02% (200 ppm)
Potato granules			0.001% (10 ppm)
Potato flakes, sweet potato flakes and dehydrated potato shreds			0.005% (50 ppm)

These limits on total product weight

FOOD STANDARDS

Enriched rice	21 CFR 137.350	BHT	0.0033% (33 ppm)
Soda water (non-alcoholic beverages)	21 CFR 165.175	BHA, BHT, TBHQ, Propyl gallate	0.02% (200 ppm) singly or in combination on fat or oil content
		Tocopherols	Amount required for intended technical effect
Frozen raw breaded shrimp	21 CFR 161.175	BHA, BHT	0.02% (200 ppm) on fat or oil content
		Tocopherols	Amount required for intended technical effect
Margarine	21 CFR 166.110	BHA, BHT, TBHQ, Propyl gallate	0.02% (200 ppm) singly or in combination on finished product (TBHQ on fat or oil)
		Tocopherols	Amount required for intended technical effect
Mixed nuts	21 CFR 164.110	BHA, BHT, TBHQ, Propyl gallate	0.02% (200 ppm) singly or in combination based on fat or oil content
		Tocopherols	Amount required for intended technical effect

1.10 Regulations

Regulations governing the usage of antioxidants vary from country to country. Various regulations of the United States, Canada and the European Economic Community are shown.

1.10.1 *United States*

Tables 1.29 and 1.30 list current regulations published by the Food and Drug Administration and the United States Department of Agriculture which permit the direct addition of BHA, BHT, TBHQ, propyl gallate and tocopherols in food.

Labelling. Under FDA and USDA regulations, the ingredients of a food must be declared on the product label in descending order of predominance. Antioxidants must be identified by their common names and the purpose for which they were added; for example, 'BHA added to help protect flavour'.

1.10.2 *Canada*

The applications for BHA, BHT and propyl gallate can be seen in Table 1.31.

1.10.3 *EEC*

EEC Council directive references for approved antioxidants are shown in Table 1.32.

1.10.4 *Other countries*

A list of countries that permit the use of one or more of the major food antioxidants can be seen in Table 1.33.

Disclaimer

Examples of antioxidant usage and application are furnished to demonstrate technical effect and do not imply that use of the antioxidant is legal. Prospective users of antioxidants should consult the appropriate regulatory agency in their country concerning the use and legality of food antioxidants.

Table 1.30 United States Department of Agriculture (USDA) regulations

Application	Food additive Regulation	Permitted antioxidant	Limitation or tolerance (percent by weight of food unless otherwise specified)
MEAT AND MEAT FOOD PRODUCTS			
Dry sausage	9 CFR 318.7	BHA, BHT, TBHQ,[1] propyl gallate	0.003% (30 ppm) singly, 0.006% (60 ppm) in combination, with no antioxidant exceeding 0.003% (30 ppm), based on total weight of finished product
Fresh pork sausage brown and serve sausage, pregrilled beef patties, and fresh sausage made from beef and pork			0.01% (100 ppm) singly, 0.02% (200 ppm) in combination, with no antioxidant exceeding 0.01% (100 ppm), based on fat or oil content of finished product
Dried meats			0.01% (100 ppm) singly or in combination based on total weight of finished product
Rendered animal fat or combination of such fat with vegetable fat	9 CFR 318.7	BHA, BHT, TBHQ,[1] propyl gallate	0.01% (100 ppm) singly, 0.02% (200 ppm) in combination with no antioxidant exceeding 0.01%
		Tocopherols	0.03% (300 ppm). (A 30% concentration of tocopherols shall be used in vegetable oils when added to products designated as 'lard' or 'rendered pork fat')
POULTRY AND POULTRY FOOD PRODUCTS			
Rendered poultry fat or combination of such fat with vegetable fat	9 CFR 381.147	BHA, BHT, TBHQ,[1] propyl gallate	0.01% (100 ppm) singly, 0.02% (200 ppm) in combination, with no antioxidant exceeding 0.01%, based on fat content
		Tocopherols	0.03% based on fat content (0.02% in combination with BHA, BHT, propyl gallate)

[1] Combination of TBHQ and propyl gallate is not lawful.

Table 1.31 Applications for BHA, BHT and propyl gallate in Canada[1]

Application	Maximum level of use (%)[2]
Fats and oils; lard; shortening	0.02
Dried breakfast cereals; dehydrated potato products	0.005
Chewing gum	0.02
Essential oils; citrus oil flavours; dry flavours	0.125
Citrus oils	0.5
Margarine	0.01[3]
Other unstandardised foods (except unstandardised meat, fish and poultry products)	0.02[3]

[1] Part IV, Table XI, Division 16 of the Canadian Food and Drug Regulations.
[2] Singly or in combination.
[3] Percentage of the fat content.

Table 1.32 EEC Council directive numbers for permitted antioxidants

Antioxidant	EEC No.
Ascorbic acid	E300
Sodium ascorbate	E301
Calcium ascorbate	E302
Ascorbyl palmitate	E304
Tocopherol extracts	E306
α-Tocopherol	E307
γ-Tocopherol	E308
δ-Tocopherol	E309
Propyl gallate	E310
Octyl gallate	E311
Dodecyl gallate	E312
BHA	E320
BHT	E321

Table 1.33 Countries where antioxidants are permitted for food use (A = approved)

Countries	Antioxidants approved for food use				
	BHA	BHT	Gallates	TBHQ	Tocopherols
Afghanistan	A	A			A
Argentina	A	A		A	A
Australia	A	A	A	A	A
Austria	A				A
Bahrain	A	A	A	A	A
Barbados				A	
Belgium	A	A			A
Brazil	A	A	A	A	A
Canada	A	A	A		A
Chile	A	A	A	A	A
China (People's Republic)	A	A	A		
China (Taiwan)	A	A	A	A	A
Colombia	A	A	A	A	
Cyprus	A	A			A
Denmark	A	A	A		A

Table 1.33 *Continued*

Countries	Antioxidants approved for food use				
	BHA	BHT	Gallates	TBHQ	Tocopherols
Ecuador	A	A			A
European Economic Community	A	A	A		A
Finland	A	A	A		A
France	A	A	A		A
Germany	A	A	A		A
Gibraltar	A	A			A
Greece			A		A
Hong Kong	A	A			A
Hungary	A	A			
Indonesia	A				
Ireland	A	A	A		A
Israel	A	A		A	A
Italy	A	A	A		A
Jamaica					A
Japan	A	A	A		A
Kenya	A		A		
Korea, South	A	A	A	A	A
Luxembourg	A	A	A		A
Malaysia	A	A	A	A	A
Malta	A	A		A	A
Mauritius	A	A	A		A
Mexico	A	A	A	A	A
Morocco				A	
Netherlands	A	A	A		A
New Zealand	A	A	A	A	A
Nigeria	A	A			
Norway	A	A	A		A
Pakistan, West	A	A			A
Panama	A	A	A	A	A
Papua New Guinea	A				A
Peru	A	A			A
Philippines	A	A	A	A	A
Portugal	A	A			
Saudi Arabia	A	A	A	A	A
Singapore	A	A	A		A
South Africa	A	A	A	A	A
Spain	A	A			A
Sweden	A	A	A		A
Switzerland	A	A	A		A
Thailand	A	A	A	A	A
Trinidad/Tobago	A	A	A		
Turkey			A	A	
United Kingdom	A	A	A		A
Uruguay	A	A			A
United States of America	A	A	A	A	A
Venezuela	A	A	A	A	A
Zambia	A	A	A		
Zimbabwe	A				

Sources:
1. Food Additive Regulations from the specific countries.
2. International Life Sciences Institute – Nutrition Foundation Antioxidant Technical Committee mongraphs on BHA, BHT, propyl gallate, TBHQ, and tocopherol.

Acknowledgements

Unless otherwise noted, the information presented in this chapter was derived from technical publications provided by Eastman Chemical Products, Inc., Kingsport, TN; and personal communication with researchers at Eastman Chemical Products.

References and further reading

Allen, J. L. and Hamilton, R. J. (eds) (1983) *Rancidity in Foods*, Applied Science Publishers, London.

Food Fats and Oils (1974) published by the Institute of Shortening and Edible Oils, Inc., August 1974.

Hudson, B. J. F. (ed.) (1990) *Food Antioxidants*, Elsevier Science Publishers Ltd., Essex.

Lundberg, W. O. (1961 and 1962) *Autoxidation and Antioxidants*, Volumes I and II, Interscience Publishers, New York.

Sherwin, E. R. (1972) Antioxidants for food fats and oils, *JAOCS* **49** (8), 468–274.

Sherwin, E. R. (1978) Oxidation and antioxidants in fat and oil processing, *JAOCS* **55** (11), 809–814.

Swern, D. (1964) *Bailey's Industrial Oil and Fat Products*, 3rd Edition, Chapter 1, *Structure and Composition of Fats and Oils*, Pages 3–53, and Chapter 2, *Reactions of Fats and Fatty Acids*, Pages 55–95, Interscience Publishers, New York.

Tenox® Food-Grade Antioxidants, ECPI Publication No. 2G-109J.

Tenox® Food-Grade Antioxidants Improve Keeping Qualities of Refined Vegetable Oils, ECPI Publication No. 2G-248.

Tenox® GT-1 and GT-2 Food-Grade Antioxidants, ECPI Publication No. 2G-263.

2 Sweeteners

S. MARIE

2.1 Introduction

From the 1950s onwards, there have been concerns about the effect of diet upon health. High carbohydrate intake, especially sugar, was implicated in the so-called 'diseases of affluence', including cardiovascular disease, obesity, diabetes mellitus and other metabolic disorders. The role of sucrose in the incidence of dental caries has also been of great concern. The United Kingdom NACNE (NACNE, 1983) and COMA (COMA, 1984 and 1989) reports recommended certain dietary changes including significant reductions in the consumption of sugar. While the contribution of dietary sugar to obesity and other diseases was inconclusive, a link between sugar consumption and the incidence of dental caries was in evidence. The concerns reflected in these and other reports have played a major role in the commercial development of a whole range of sucrose substitutes, otherwise known as sweeteners.

The properties sought in a sucrose replacement may be summarised as:

1. The same taste and functional characteristics as sucrose.
2. Low caloric density on a sweetness equivalency basis.
3. Non-cariogenicity.
4. Metabolised normally or excreted unchanged.
5. No allergenic, mutagenic, carcinogenic or other toxic effects in the body.
6. Chemical and thermal stability.
7. Compatibility with other food ingredients.
8. Economically competitive with existing sweeteners.

Sweeteners may be nutritive, as are the hydrogenated sugars, also known as sugar alcohols or polyols, or non-nutritive, as are the intense sweeteners. They can be synthesised or extracted from natural sources. Intense sweeteners contribute no bulk, viscosity or texture to foods and beverages, and must be mixed with nutritive sweeteners or some other bulking agent when these properties are required.

The development of a wide range of sweeteners has advantages for the food industry. Because of the idiosyncratic properties of individual sweeteners, one may be better suited to a specific application than others.

Furthermore, there are possibilities for combining two or more sweeteners in a product. Improved safety is one benefit of mixtures, since the required quantity of each component sweetener is then reduced. In addition, mixtures of sweeteners provide the developer with many more opportunities, than with single sweeteners, for optimising such features as the quality and stability of the taste profile of the final product. Mixtures of sweeteners can be tailored to the product, by their mutual complementation and compensation of characteristics. For example, saccharin may be combined with aspartame, resulting in a mixture that has greater stability then aspartame and masks the after-taste of saccharin, with the bonus of synergism for sweetness. In products where bulking and texturing is required, the calorific value can be reduced, while sweetness is maintained, by combining a sugar alcohol with an intense sweetener. The lingering, menthol-like sweet taste of thaumatin may be desirable for oral-care products to achieve a long-lasting fresh taste in the mouth, and the delayed onset of thaumatin-taste can be compensated for by its combination with another sweetener providing instant sweetness.

The ideal replacement for sucrose has yet to be found, and the volume of worldwide applications for patent registrations of sweet substances is testimony to the continuing quest to achieve this goal. Few, however, survive the rigours and expense of the full gamut of safety evaluations now required for entry into the market place.

The sweeteners discussed here are the polyols, including sorbitol, mannitol, xylitol, maltitol, lactitol, isomalt, and Lycasin®, a hydrogenated glucose syrup; the intense sweeteners, including saccharin, cyclamate, aspartame, acesulfame-K, thaumatin, stevioside and neohesperadin dihydrochalcone (NHDC); and some new sweeteners currently being developed, including sucralose and RTI-001. The sensory and physical properties of these sweeteners are detailed in Tables 2.1–2.3. Information of the general background of individual, or groups, of sweeteners, together with details of mixtures, health and safety, regulations and applications, is provided in the text.

2.2 Polyols

2.2.1 General

The polyol sweeteners are sugar alcohols produced by the hydrogenation of sugars and syrups, with the aid of a catalyst, usually Raney nickel. These include *sorbitol, mannitol, xylitol, maltitol* and *lactitol*, derived from glucose, mannose, xylose, maltose, and lactose, respectively; *isomalt*, an equimolar mixture of glucopyranosyl-sorbitol and glucopyranosyl-mannitol resulting from the hydrogenation of isomaltulose (also known as palatinose); and *hydrogenated glucose syrups* (*HGS*), such as *Lycasin*® *80/55*, derived from starch hydrolysates.

The first hydrogenated sugar to have been manufactured was *sorbitol* in the 1930s. This was followed by *mannitol, HGS* including *Lycasin*®, and *xylitol* in the early 1970s. *Isomalt, maltitol* and *lactitol* were more recent developments.

A major sorbitol producer, Roquette Frères of France, acquired the patent rights to *Lycasin*® in 1975. *Isomalt* is marketed by Suddeutsche Zucher AG, West Germany, as *Palatinit*®, and by Tate and Lyle plc, UK, as *Lylose*®. *Lactitol dihydrate* was developed by CC Biochem of The Netherlands, and is sold exclusively by Philpot Dairy Products in the UK under the brand name *Lacty*® (Anon, 1989d). *Xylitol* is marketed by Finnsugar Xyrofin. *Maltitol* is available as *Malbit*®, with <90% maltitol and >5% maltotriitol, and Roquettes Frères applied for a French patent in 1987 for compressible *maltitol* (>85%) powder for use as a sweetener (Anon, 1987a).

2.2.2 Properties

Physical. Sugar alcohols are functionally similar to sucrose, and are bulking agents. *Sorbitol, mannitol, xylitol* are naturally occurring, like sucrose, thus promoting consumer appeal. There are both technical and physiological benefits of these conversions. The technical benefits include increased chemical stability and affinity for water, without altering the sweetening power, and a reduced tendency to crystallise. The physiological benefits are that the sugar alcohols have low cariogenicity. They are suitable for inclusion in products for diabetics because their metabolism is insulin-independent, and some are low in energy because of their malabsorption in the intestine (Hough, 1979). *Lycasin* was developed as a product to replace sucrose that was not harmful to teeth.

There are specific properties of the sugar alcohols (summarised in Table 2.1) that distinguish them apart, and bestow upon them particular advantages and disadvantages in their applications. For instance, *sorbitol* has special properties of high viscosity, humectancy and crystal form. It is readily soluble in water, but virtually insoluble in all organic solvents, except alcohol (Hough, 1979). *Mannitol* has low solubility in water, but is useful as an anti-adhesion agent, inhibiting the crystallisation of other polyalcohols in the manufacture of chewing gum (Sicard and Leroy, 1983). *Maltitol* is highly hygroscopic, inhibiting crystallisation; has excellent heat stability, with no loss of colour during boiling (Fabry, 1987); and has low fermentability by common moulds and bacteria (Grenby, 1983). *Lactitol* has good solubility in water, that increases with temperature. At 25°C, 150 g of *lactitol monohydrate* or 140 g of *lactitol dihydrate* (*Lacty*®) will dissolve in 100 ml water. Its white, crystalline, solid form melts at about 120°C for the *lactitol monohydrate*, and at about 75°C for the *dihydrate*, and is partly converted into lactitan, *sorbitol* and lower polyols when heated at 179–240°C (den Uyl, 1987). *Lactitol* has some useful parallels with sucrose. Its viscosity in solution is equal to that of sucrose, weight for weight, and it lowers the freezing points of solutions, in the same way as sucrose – important in making ice-cream. Decomposition is a function of temperature and acidity, but *lactitol solutions* have excellent storage stability in the pH range 3.0–7.5 and at temperatures of up to 60°C. While no detectable decomposition of 10% *lactitol solutions* was observed under these conditions after 1 month, there was 15% decomposition at pH 3.0 after 2 months, but none at 105°C at pH 12.0 (den Uyl, 1987).

Lycasin® *80/55* is a clear, colourless, odourless syrup, also available in a powder form, that is very stable, chemically inert, and has technical properties that permit its replacement of sucrose in many food products, especially confectionery.

Isomalt is white and odourless, with a crystalline form. Its solubility in water is a function of temperature, with about 25% solubility at 20°C and 55% at 60°C. Solubility decreases linearly, though, with the addition of alcohol (Strater, 1986). *Isomalt* is highly stable with respect to chemical and microbial breakdown. It has no Maillard reaction so that browning inhibitors are unnecessary. *Isomalt* melts within the range 145–150°C, and changes colour only slightly when held at 170°C for 60 minutes. There are no further colour changes after the first hour, in contrast to sucrose solutions, which suffer exponential increases in colour (Strater, 1986). Its viscosity in aqueous solution is comparable to that of sucrose solutions.

Sensory. *Xylitol* has about the same sweetness as sucrose, weight for weight, with slight variations due to conditions of temperature, pH and concentration (Anon, 1986a). The other sugar alcohols are less sweet (Table 2.3) and need supplementation with intense sweeteners to be comparable to sucrose. Reports of the sweetness of *Lycasin*® *80/55* vary between 55% and 75% that of sucrose.

Sugar alcohols add texture and mouthfeel properties to foods and drinks since they are bulking agents. Many of the sugar alcohols, *xylitol* and *sorbitol* in particular, impart a cooling sensation in the mouth because they absorb heat as they dissolve. Isomalt, however, does not possess this characteristic. All have pleasant, sweet taste profiles with no after-taste.

2.2.3 *Mixtures*

Sugarless chewing gums typically contain *sorbitol* and/or *mannitol* as sugar substitutes, and the reduced sweetness is supplemented with the addition of intense sweeteners. Saccharin is commonly used, because sufficiently small concentrations are needed for the saccharin taste limitation to be imperceptible. *Sorbitol* is also combined with intense sweeteners in products for diabetics.

Xylitol is used, especially in Europe, alone or with other polyols or polydextrose in sugarless confectionery products. Some advantages of the bulking agent, polydexterose, are that it has low laxative properties, low calorific value (1 kcal g^{-1}) and is tolerated by diabetics. It is, however, low in sweetness.

It is claimed that no intense sweeteners are required to supplement *Malbit*® in order to obtain the sweetness and flavour release properties close to those of sucrose-sweetened products (Fabry, 1987). It is often desirable, by contrast, that the sweetness of *lactitol*-containing products be increased, preferably by the addition of aspartame or acesulfame-K. A 10% *lactitol solution* containing 0.03% aspartame or acesulfame-K has the equivalent sweetness of a 10% sucrose solution (den Uyl, 1987).

Isomalt is synergistic with other sugar alcohols, such as *sorbitol, xylitol* and *HGS*, and with intense sweeteners, such as saccharin and aspartame. Metallic after-tastes of intense sweeteners are masked in such mixtures. *Isomalt* can also be mixed with the low-calorie bulking agent, polydextrose, in the production of calorie-reduced foods (Mackay, 1987).

Lysasin® has an agreeable sweet taste that needs no supplementation in confectionery with intense sweeteners (Sicard and Leroy, 1983). However, it may be added to products with other polyols, such as *sorbitol* and *isomalt*, as a crystallisation inhibitor.

2.2.4 *Health and safety*

The most attractive feature of the polyol sweeteners is that they are suitable as sugar substitutes for caries prevention. *Xylitol* is outstanding in this regard. Long-term field trials have been carried out, such as the Turku studies in Finland in the 1970s, and those commissioned by the WHO in the mid-1980s with caries-prone children in Hungary and French Polynesia. The results clearly demonstrated the caries inhibiting effect of *xylitol* when consumed in small quantities (14–20 g) as an addition to the daily diet. Subsequent studies have confirmed the protective effect of *xylitol*-sweetened chewing gum, although the mechanism by which *xylitol* inhibits caries development is unknown (Pepper and Olinger, 1988).

The close structural similarity between many bulk sweeteners and normal carbohydrates has led to limited applications for toxicological assessments. The absence of any adverse effects, though, i.e. complete food safety, has been demonstrated for *xylitol*, and for *isomalt*, the components of which are found in the body anyway (Ziesenitz and Siebert, 1987). However, a limitation of the polyols is their laxative effect at high doses, and warning labels on products are required in some countries.

Sorbitol has been used since the 1920s for sweetening foods for diabetics, since its metabolism causes only an insignificant rise in blood glucose. Similarly, *xylitol* (Sicard, 1982), *lactitol* (den Uyl, 1987) and *isomalt* (Strater, 1986) are metabolised with no significant changes of blood glucose and insulin, and so are suitable for inclusion in the diets of diabetics. *Mannitol* is less desirable for diabetics because of its low laxative threshold (Table 2.3). *HGS* and *maltitol* are also disadvantageous because glucose is a breakdown product in the gut (Dwivedi, 1987).

The malabsorption of some of the polyols, such as *lactitol*, gives rise to a beneficial physiological side effect. Such unabsorbed polyols act as dietary fibre, being fermented by the microflora of the large intestine and contributing to faecal mass (Booy, 1987).

A limitation of sugar alcohols is their laxative effect when consumed in large doses. However, the EEC Scientific Committee on Food (SCF) advised that 20 g per person per day of polyols is unlikely to cause undesirable laxative symptoms (Anon, 1984). In some countries, product labels may be required to carry an appropriate warning, depending upon the amounts involved.

2.2.5 Regulations

Isomalt, mannitol, sorbitol, sorbitol syrup, xylitol and *hydrogenated glucose syrups* were among the 12 sweeteners listed as permissible for food use in the UK (Statutory Instruments, 1983). *Lactitol* was added in 1988 (Sweeteners in Foods (Scotland) Amendment Regs., 1988).

Palatinit Isomalt was under examination earlier this year (Anon, 1989c) by the National Health and Medical Research Council for inclusion in Australian Food Standards.

Xylitol is permitted in more than 40 countries for use in foods and other products, including the EEC, North America and Scandinavia (Anon, 1989d). It was given FDA and WHO/FAO clearance in 1978 for sweetening of Special Dietary Foods, but not GRAS (Generally Recognized As Safe) status. *Sorbitol* and *Lycasin*®, however, have been affirmed as GRAS (Mackay, 1979).

Maltitol is permitted in certain foods in Japan, and HGS in Switzerland, The Netherlands and Scandinavia. These sweeteners are not permitted for food use, though, in the US. *Sorbitol* and *mannitol* are listed as permitted sweeteners or food additives, subject to certain restrictions, in the US, South Africa, UK, Belgium, Denmark, Greece, Spain, France, Germany, Switzerland, Sweden, Japan, Australia and Canada, and *sorbitol* additionally in The Netherlands, Italy, Norway, Finland and Brazil.

JECFA allocated an ADI 'not specified' for *lactitol*, in 1983, and for *isomalt* and *HGS* in 1985, with the additional comment that levels should be appropriate in consideration of the known laxative effect of polyols (Joint FAO/WHO, 1987).

2.2.6 *Applications*

Sugar alcohols are indispensable in products marketed as low in cariogenic-ity, and for diabetics, when a bulking function is required.

Economics is an important consideration in applications. *Xylitol* is a relatively expensive polyol (Table 2.1; Dwivedi, 1986) but this could be counterbalanced by its wide consumer appeal on account of its dental advantages and its status as a naturally occurring compound. *Mannitol* has also been a very expensive sweetener (Table 2.1; Dwivedi, 1986) because of the purification processes required to separate it from *sorbitol* and other contaminants in its manufacture, but methods of reducing these production costs are being researched. *Mannitol* is mainly used in the crystalline form in sugar-free chewing gum, and also in chewable pharmaceutical products as it is inert (as are other sugar alcohols) to most drug components. Its application in products such as soft drinks, ice-cream and confectionery is restricted by its low solubility.

Sorbitol is an attractive sweetener for manufacturers who wish to make use of its special properties since it needs no special handling. For instance, it is used in fondants, fudges, marshmallows and caramels to retard sucrose crystallisation and thus to retain freshness and flavour. It acts as a humectant and anticaking agent in baked goods. *Sorbitol*, supplemented by an intense sweetener, is used to sweeten diabetic products such as table-top sweeteners, preserves, jellies, and confectionery. It may totally replace sucrose in chocolate and ice-cream for diabetics, although the final products are distinguishable (Hough, 1979).

Xylitol is used mostly as a sweetener in sugarless chewing gum. It also has potential in tableted products such as mints and children's chewable vitamin tablets. This can be in a 50:50 combination with sorbitol (the gamma form of tableting grade is preferred), both milled, blended and compressed. However, this combination suffers from poor flow properties, and the direct compression of *Xylitol DC*, containing not more than 4% *sorbitol* may be preferable. This method, using a force of 20 kN to produce tablets of 15 mm diameter, results in a satisfactory hardness, since the melted *xylitol* recrystallises during overnight storage (Pepper and Olinger, 1988). In chocolate, *xylitol* is a good choice as a substitute for sucrose because of its equi-sweetness, but its high relative cost may be prohibitive. Conching can take place at temperatures of up to 55°C. *Xylitol* is successfully combined with *sorbitol* to provide the syrup phase in fondants, and is exceptionally good with mint and chocolate flavours. Its use in pectin and gelatine jellies produces high-quality products, but, since *xylitol* reduces gel strength, extra gelling agent is required (Pepper and Olinger, 1988).

Maltitol aids moisture retention in baked goods, and is regarded as suitable for carbonated beverages, canned fruits (Hough, 1979) and confec-tionery, especially gloss coatings (Grenby, 1983). *Malbit*® is available in both crystalline and liquid forms, with sweetnesses, relative to sucrose, of 0.8–0.9 and 0.6 respectively (Table 2.3). Due to its physiological, organoleptic and technical properties, *Malbit*® can be used in the development of a new generation of health and speciality products, including dietary, diabetic, tooth-protective and slimming products. It has been used in products in Japan for ten years, and is now on the market in some European and Asian countries in dark and milk chocolate, hard-boiled candies, soft caramels, toffees and chewy fruits, fruit pastilles and liquorice gums based on gum arabic, gelatin gums and jellies, chewing gums and bubble gums, chocolate dragees using panning technology, other types of confectionery including sugarless tablets and muesli bars, jams, and ice-cream (Fabry, 1987).

Lactitol can replace sucrose as a sweet-tasting texturising or bulking agent in a variety of applications with equal palatibility and no aftertaste (den Uyl, 1987). According to Philpot Dairy Products, *Lacty*® can be used in bakery products, confectionery, ice-cream, jams and marmalades and table-top sweeteners (Anon, 1989d). Because of its low hygroscopicity, *lactitol* is particularly suitable, in conjunction with its low caloric value, for use as a bulking agent for intense sweeteners in table-top use, and for biscuit making since crispness is maintained. It is also successfully used as a surface dusting for confectionery. *Lactitol* is suitable, also, for inclusion in low-calorie and sugarless products such as chewing gum, fruit gums and pastilles, chocolate, instant beverages and jams. The inclusion in the product of a crystallisation inhibitor, such as *Lycasin*, may be necessary.

Isomalt is a suitable ingredient for confectionery, baking and soft drinks. Its ready crystallisation simplifies the coating of hard-boiled and chewable candies, and can be used to enhance the shelf-life of hygroscopic products. Because of the high percentage of solids dissolved in the aqueous phase, *isomalt* can also be used as a melt for the manufacture of soft caramels, chewing gums and soft candies, probably with the addition of a crystallisation inhibitor, such as *HGS* (Strater, 1986).

Lycasin has the same viscosity, colour and shelf life as ordinary starch syrups, and so can perform the same function in sugar-free products, such as confectionery and jam (Rockstrom, 1980). *Lycasin* is highly hygroscopic due to its *maltitol* content. Because of this characteristic, *Lycasin*® 80/55 is useful in the manufacture of liquid-centre confectionery, and has an anticrystallising effect on other product ingredients, such as *sorbitol* in chewing gum (Sicard, 1982).

2.3 Saccharin

2.3.1 *General*

Saccharin was accidentally discovered in 1879 by Fahlberg and Remsen, and manufactured five years later. It was first used as an antiseptic and preservative, but as a sweetener since 1900. Saccharin is synthesised com-

Table 2.1 Properties of sucrose and polyols

	Sucrose	Sorbitol	Mannitol	Xylitol	Malbit®	Lactitol	Isomalt	Lycasin®
Sensory								
Sweetness intensity	1	0.6	0.6	1	0.6–0.9	0.35	0.45	0.55–0.75
Mouth-cooling effect	None	High	Low	High	Low	Low	None	Not studied
Physiological								
Energy (kcal g^{-1})	4	4	<4	<4	2	2	2	
Cariogenic potential	High	Low	Low	None	None	None	None	Low
Suitability for diabetics	None	High	Low	High	Low	High	High	Low
Laxative at (g day^{-1})	No effect	50–75	20	50–70	50	70–80	20–30	30–50
Physical								
Molecular weight	42	182	182	152	344.47 (maltitol)	362 monohydrate, 380 dihydrate	368	340+
Hygroscopicity	High	High (solutions) Low (powder)	Low	High	High	Low	Low	High
Browning reaction	Yes	No	No	No	No	No	No	No
Solubility in water (g (100 ml)$^{-1}$) room temp.	High (66)	High (75)	Low (18)	High (63)	Medium	High (149 monohydrate, 140 dihydrate at 25°C)	Low (25)	Supplied as syrup
Melting point		96–97	165–168	93–94.5	135–140	115–125 monohydrate, 70–80 dihydrate	145–150	
Stability	Stable in neutral pH	Stable to heat, chemically unreactive	Chemically stable	Chemically stable	Chemically and thermally stable	Solutions: good at pH 3.0–7.5 and at temps. <60°C for 1 month	Resistant to chemical and microbial breakdown	Stable, chemically inert
Other								
Price ratio (1986)	1	1.4 liquid 2.3 crystalline	4.0 crystalline	10.0 crystalline				
ADI mg (kg body weight)$^{-1}$		'None specified'		'None specified'		'Not specified'	'Not specified'	'Not specified'

Sources: Sicard and Leroy (1983); von Rymon Lipinski (1987); Grenby (1983); Pepper and Olinger (1988); den Uyl (1987); Hough (1979); Ziesenitz and Siebert (1987); Dodson and Pepper (1985); Sicard (1982); Anon (1988); Joint FAO/WHO (1987); Booy (1987); Fabry (1987); Dwivedi (1986); von Hertzen and Lindqvist (1980).

mercially from toluene, and has a chemical formula of $C_7H_5NO_3S$. It is usually available as the sodium salt and sometimes as the calcium salt. Other saccharin salts are not commercially available, although reputedly sweet (Walter and Mitchell, 1986).

Saccharin is the most widely used sugar substitute in the world, probably because of its high stability and low cost, and the only available intense sweetener in some countries for many years. It has been called the 'pioneer' sweetener, paving the way for a variety of low-calorie products (Bakal, 1987), but has the disadvantage of a bitter, metallic after-taste.

2.3.2 *Mixtures*

Mixtures with saccharin have been made for three reasons: to mask the unpleasant taste characteristics of saccharin; to provide bulk as well as sweetness; and to take advantage of synergy for sweetness (Table 2.3).

Many mixtures, thought to improve the taste of saccharin, have been patented. These have included combinations with other sweeteners, notably cyclamates and aspartame, as well as with such ingredients as cream of tartar, glucono-δ-lactone, sodium gluconate, glycols, gentian root, maltol, pectin, lemon flavour, ribonucleotides and adipic, aldohexuronic and citric acids. For example, cyclamate/saccharin combinations at a ratio of about 3:1 were most successful, when both were permitted sweeteners, with high consumer acceptance and providing sugar-like sweetness in beverages. Since cyclamate was banned in the US in 1969, it has been replaced by calcium chloride in combination with cornstarch hydrolysate, lactose, sucrose, tartrates, and fructose with gluconate salts in combination with saccharin (Bakal, 1983; Walter and Mitchell, 1986).

Saccharin is synergistic with cyclamate and with aspartame.

2.3.3 *Health and safety*

Saccharin is not metabolised in the body, but is excreted unchanged. Although bladder tumours have been associated with saccharin intake in rats, extensive research on human populations has established no such association. Research included three major studies with diabetics, who consume greater amounts of saccharin than the general public, but no increased risk of cancer was in evidence with this group (Walter and Mitchell, 1986).

Saccharin has been assigned an ADI of $2.5\,mg\,kg^{-1}$ of body weight. However, there are fears presently that this level may be exceeded by some sectors of the population, and the Ministry of Agriculture, Fisheries and Food (MAFF) in the UK have called for an investigation (Anon, 1989b).

Recent research indicates that saccharin has anticariogenic properties, rather than non-cariogenic properties as previously supposed (Linke, 1987).

2.3.4 *Regulations*

Saccharin is used in more than 80 countries, but its approval in the US has had a stormy history. In 1977 the FDA proposed to ban it because of the discovery of bladder tumours in rats fed on high doses of saccharin. However the ban was suspended due to its extreme unpopularity, while further investigations were completed. In July 1987, the Judicial Review of the Code of Federal Regulation with regard to Food Additives (CFR, 1989) extended the moratorium on saccharin until 1 May 1992. Despite the caution of the FDA, reviews by other regulatory bodies have approved the use of saccharin. These have been the Food Additives and Contaminants Committee (FACC) of Great Britain in 1982, the Joint Expert Committee on Food Additives (JECFA) in 1984, and the Scientific Committee for Food of the Commission of the European Economic Communities (EEC) in 1984 (Walter and Mitchell, 1986).

2.3.5 *Applications*

Saccharin has a wide range of applications due to its high stability, nil calorific value, non-cariogenicity, and its low cost. A prime asset of saccharin is its high stability in a wide range of products even under extreme conditions of processing. It is the only approved sweetener which can withstand heating, baking and high-acid media (Bakal, 1987), and is one-twentieth the price of sugar in terms of sweetness equivalency.

It has been used in soft drinks, candies and preserves, salad dressings, low-calorie gelatine desserts, and combined with bulk sweeteners in baking for sugar-reduced products. It has been produced as a variety of table-top preparations either as a single sweetener, in tablet and liquid form, or in combination with other sweeteners, and incorporated into chewing gum on its own or combined with sorbitol or aspartame. Saccharin is also a popular choice in oral-hygiene products such as toothpaste and mouthwashes.

The future prospects for saccharin in the market place are unclear because of consumer concerns about its safety and because of the present availability of alternative sweeteners such as aspartame and acesulfame-K. Its survival may be restricted largely to combinations with other sweeteners.

2.4 Cyclamate

2.4.1 *General*

Cyclamate (N-cyclohexyl-sulphamic acid) was discovered in 1937 by Michael Sveda of Abbott Laboratories, Chicago. It was used as a sweetener from the mid-1950s, becoming the dominant artificial sweetener in the 1960s in the form of its sodium and calcium salts. It was the major factor in launching the diet segment of the food and beverage industries. However, it lost GRAS status in 1969, was banned in 1970 in the USA, and soon after in the UK and other countries. Its ban resulted in the deterioration of taste profiles of soft drinks, in particular, and an incentive to develop new sweeteners. Cyclamate is still permitted in some applications, however, in some countries including Spain, Germany and Switzerland (Kasperson and Primack, 1986).

2.4.2 Mixtures

Cyclamate has only one tenth the sweetness of the equivalent weight of saccharin. However, it was found in the 1950s that their combination in the ratio 10:1, on a sweetness equivalency basis, produced a most desirable sweetness (Miller, 1987). Cyclamate masked the after-taste of saccharin, and the low sweetness of cyclamate was boosted by saccharin and by the synergy of their mixture (Table 2.3). This combination became the first commercial multiple sweetener, and was used in the 1960s in diet soft drinks, table-top sweeteners, low-calorie frozen desserts, salad dressings, jams, jellies, and other products (Gelardi, 1987). It remains a widely used sweetener mixture, even though cyclamate is now banned in some countries (Lindley, in press).

Combinations of cyclamate with aspartame, and with aspartame and saccharin together, have also been found to improve stability and to give good taste profiles in table-top sweeteners, diet soft drinks, dry beverage mixes and chewing gum (Gelardi, 1987). Cyclamate is also synergistic with sucrose (Table 2.3).

2.4.3 Health and safety

Cyclamate has no calories and is non-cariogenic, but there have been fears concerning its toxicity. Early studies concluded that the compound was poorly absorbed in the gut, and excreted unchanged, thereby excluding any undesirable metabolic effects or by-products. However, later evidence contested this finding. In an experiment with rats (Price et al., 1969), fed daily for a lifetime on high doses of cyclamate and saccharin in a 10:1 ratio, it was found that cyclamate was metabolised to a product called cyclohexylamine, and this metabolite became implicated in the occurrence of bladder tumours appearing after two years. Although the evidence against cyclamate was not 100% conclusive, there were also fears about other effects on genetic material, leading to its withdrawal in some countries.

2.4.4 Regulations

Cyclamate was approved as a food additive by the FDA in 1949, and achieved GRAS status in 1958. However, it was banned in the US with effect from 1970 (Federal Register, 1969) because of its association with tumours in rats. Bans in the UK, Canada and Japan followed, while restrictions were made in several European countries.

Abbott Laboratories petitioned the FDA in 1973 and in 1980 to reapprove cyclamate on account of the numerous studies that failed to confirm the carcinogenicity of cyclamate or its metabolite. In 1985, the National Academy of Sciences (NAS) supported the conclusion of the 1984 Cancer Assessment Committee that cyclamate is not a carcinogen. In 1986 the FDA arranged further toxicological tests of food additives, including cyclamate. However, the ban on cyclamate and its derivatives for food uses was reaffirmed in April 1989 (CFR, 1989).

In 1982, a review of sweeteners in the UK recommended a continued prohibition of cyclamate due to its unknown effects on man, but JECFA, in the same year, replaced a temporary approval of cyclamate with a full one and an increased ADI of 11 mg kg^{-1} of body weight (Higginbotham, 1983).

2.4.5 Applications

Cyclamate has the benefits of good taste and low cost but, where it is permitted, the quantities required on account of its low sweetness are likely to be in excess of ADI amounts. Further applications, then, are likely to be in mixtures of sweeteners.

2.5 Aspartame

2.5.1 General

The sweet taste of the compound, aspartame, was discovered accidentally in 1965 by James Schlatter, while synthesising a product for ulcer therapy. Aspartame is a dipeptide methyl ester, composed of the two amino acids, phenylalanine and aspartic acid. It is marketed by G.D. Searle as Nutra-sweet®, as Equal® and as a tablet sweetener called Canderel®, and by the Holland Sweetener Company of The Netherlands as Sanecta®. It has a very agreeable sweet taste but is unstable under certain conditions (Table 2.2).

2.5.2 Mixtures

Mixing aspartame with other sweeteners has the advantages over its use as a single sweetener of improving processing and shelf stability, while producing a balanced taste. Furthermore, mixtures of aspartame with acesulfame-K, or with sodium saccharin, sodium cyclamate, glucose or sucrose, are synergistic (Table 2.3), having the added advantage of cost reduction.

Table 2.2 Properties of intense sweeteners

	Saccharin	Cyclamate	Aspartame	Acesulfame-K
Source	Synthetic	Synthetic	Synthetic	Synthetic
Discovery	1879	1937	1965	1967
Appearance	White, crystalline powder	White, crystalline powder	White, crystalline powder	White, crystalline powder
Molecular weight	205 Na salt		294	201
Solubility (room temp.)	82% Na salt 67% Ca salt	Na and Ca salts readily soluble in water	38% water; 0.4% ethanol at 25°C; not soluble in fats and oils	31% water (100% at 100°C); 0.1% ethanol; >30% dimethylsulphoxide
Stability: pH	Stable in range 2–7	Stable in range 2–7 at normal process temperatures	After 36 days 50–60% degraded at pH 3.5, fully hydrolised at pH 7.4 by day 9	Stable for several months at pH 3 or more
Temperature	Unchanged after 1 h at 150°C, pH 3.3–8.0	Temperature-dependent decomposition; in aqueous solution, pH 2.1, hydrolises producing 350 mg cyclohexylamine per kg cyclamate at 30°C, 500 mg kg^{-1} at 44°C, after 40 days	Unstable at high temperatures	Stable to pasteurisation and sterilisation if pH >3 and to baking at temps. >200°C
Storage	Stable in dry form for several years	Stable in tablet form for several years; in aqueous solution hydrolises slowly to sulphuric acid and cyclohexylamine	Stable in dry form; unstable in aqueous solution, 50% degraded after 36 days	Shelf-life of >5 years in solid form; no hydrolysis of sterilised solution stored one month at 40°C
Other		Decomposition accelerated in presence of amino acids and water-sol. vitamins at elevated temps.		
Melt point	229–230°C		246–247°C	Decomposition at 225°C on slow heating
Browning reaction	None	None	None	None
Cariogenic	Anti-cariogenic	Non-cariogenic	Anti-cariogenic	Non-cariogenic
ADI mg kg^{-1} b.w.	2.5 JECFA	Banned US, 1970 11.0 JECFA 1982	40.0 JECFA	9.0 JECFA 15.0 FDA
Cost vs sucrose (equi-sweet)	Lower	Lower	Higher	Same

Stevioside	Thaumatin	NHDC	Sucralose	RTI-001
Natural extract	Natural extract	Natural or synthetic	Derived from sucrose	Synthetic
1905	1972	Late 1950s	1976	1981
White powder (90% pure)		Crystalline powder	White, crystalline powder	
805	21 000		397.64	
>40% in water, insoluble in ethanol	60% in water, good solubility in ethyl and isopropyl alcohols, glycerol, propylene glycol and higher polyols such as sorbitol. Insoluble in ether, acetone, toluene, and triacetin.	Low solubility, 1.2%, in water	28% at 20°C in water, soluble in lower alcohols and other polar solvents	
Stable in range 3–9	Stable in range 1–9 at ambient temps.	Unstable at low pH	Stable in solution at low pH	After 36 days: no change at pH 3.5, 30–40% degraded in water, 70% degraded at pH 7.4
Stable at room temperature, withstands 100°C 1 h, pH 3–9	Stable to heat in range pH 2.7–6.0, optimum pH 2.8–3.5, withstands 100°C at pH <5.5 for several hours	Adequate stability in formulations; buffered solutions pH 2–1 stable 8 h, at 100°C	Stable in solution at high temperatures as sucrose	Thermostable
Stable at room temp. in citric and phosphoric acidified beverages for 3 and 5 months respectively	Indefinite in dry form; several years for chemically preserved solutions at ambient temps.	Occasional yellow discoloration in aqueous solution	Several years in liquids	Twice the shelf-life of aspartame in acidic solutions
Stable in presence of salt	Taste reduced by mono- and divalent salts, increased by tri-valent salts; denatured by metaphosphoric and phytic acids at pH 2.9; loss of sweetness with xanthan, CMC, pectin and alginate; incompatible with carrageenans		Resistant to enzymic hydrolysis	
196–198°C	172–174°C	172–174°C		
None			None with proteins, gums, tannins and other carbo-hydrates	
Non-cariogenic	Non-cariogenic	Non-cariogenic	Non-cariogenic	Low cariogenic potential
—	'Not specified'			
Lower	Higher			

Sources: Walter and Mitchell (1986); Dodson and Pepper (1985); Bakal (1983, 1987); Lindley (1983); Higginbotham (1979, 1983, 1986); Ripper *et al.* (1986); Sunett® Technical Brochure; Andres (1987); Bakal and O'Brien Nabors (1986); Kasperson and Primack (1986); Joint FAO/WHO (1987); Horowitz and Gentili (1986); von Rymon Lipinski (1986).

2.5.3 *Health and safety*

Aspartame is one of the most thoroughly tested food additives. The safety of aspartame's component amino acids, aspartic acid and phenylalanine, and of its metabolite, methanol, has been questioned. However, toxicity is always dose-related and substantial safety margins have been reported with respect to amounts likely to be consumed in the human diet (Ripper *et al.*, 1986). Garriga and Metcalfe (1988, cited by Anon, 1989a) concluded, from an analysis of adverse reactions and clinical data, that aspartame is remarkably safe. However, appropriate warnings on product packaging are necessary to alert sufferers of phenylketonuria, since there is a need to control the amount of phenylalanine in their diets.

2.5.4 *Regulations*

Aspartame achieved FDA approval in 1974 for use as a sweetener, flavour enhancer and as an ingredient in some dry food products. However, objections to the approval led to its suspension pending authentication of the safety studies. In 1981, the FDA reinstated the original approval and, in 1983, granted permission for aspartame's use in carbonated beverages and carbonated beverage syrups. A further amendment extended its use to chewable multivitamin tablets (Ripper *et al.*, 1986). By 1987, four more approved food categories had been added. These were frozen juice drinks, frozen novelties on a stick, tea beverages and breath mints (Andres, 1987).

Aspartame is now a permitted food and beverage additive and/or table-top sweetener in more than 50 countries (Ripper *et al.*, 1986).

2.5.5 *Applications*

The availability of aspartame has contributed greatly to the upsurge in low-calorie products in the market place. Its good taste, as well as its make-up from food-associated amino acids, have promoted its acceptability. Its popularity has been in spite of the high cost of aspartame, relative to other sweeteners and sucrose. However, the price can be expected to be reduced after Searle's US patent expires in 1992, because of the entry of competing manufacturers into the market place. One poised to do so is a Japanese–Dutch firm called Toyo–DSM Aspartame VOF (Anon, 1986b).

Aspartame blends well with other food flavours, but interacts with them differently than does sucrose, and so should not be used as a simple substitution for sucrose in formulations (Mazur and Ripper, 1979). It also has flavour-enhancing properties particularly with citrus fruit drinks.

The unstable character of aspartame has placed limitations upon its applications. However, these have been overcome or accommodated to some extent. For instance, although conditions of pH, temperature and moisture cause loss of aspartame due to decomposition, together with reduced sweetness, it has been shown that overall acceptability of certain carbonated, soft drinks is not affected pro rata, but remains high over a wide range of concentrations (Ripper *et al.*, 1986). Also, processing of foods containing aspartame at high temperatures is now possible with HTST (high temperature/short time) technology. Dairy products can be pasteurised by this method, and applications in the bakery industry are no longer impossible (Andres, 1987).

2.6 Acesulfame-K

2.6.1 *General*

Acesulfame-K is a potassium salt derived from acetoacetic acid, with a chemical formula of $C_4H_4NO_4KS$ and a molar mass of 201.2 (Sunett[®] Technical Brochure). This sweet-tasting compound was discovered by accident by the employees of Hoechst AG in 1967, and is currently marketed by Hoechst under the trade name of Sunett[®].

2.6.2 *Mixtures*

Mixtures of sweeteners may be advantageous for improving the taste profile, or for economic reasons where there is synergism. Although acesulfame-K can be used alone in foodstuffs, there are practical benefits of combining it with bulk sweeteners in some applications. Combinations of acesulfame-K with sorbitol in a ratio of 1:150–200, with sucrose in a ratio of 1:100–150, with isomalt in a ratio of 1:250–300, and with maltitol in a ratio of 1:150 provide mixtures with a 1:1 ratio on a sweetness basis that appear to give the best taste profiles (von Rymon Lipinski, 1985). Acesulfame-K is said to round up the weak sweetness of sorbitol, and provide a more fully developed taste (von Rymon Lipinski, 1982).

Acesulfame-K produces a pronounced synergistic effect of sweetness intensity of up to 30% or more in combination with cyclamate and with aspartame, but a barely noticeable effect with saccharin (Table 2.3). The most favourable sensory properties were observed using combinations having inverse ratios of the components' sweetness intensities, e.g. 1:1 by weight for acesulfame-K and aspartame, and 1:5 by weight for acesulfame-K and sodium cyclamate (von Rymon Lipinski, 1986). Acesulfame-K is also synergistic with the nutritive sweeteners sorbitol, isomalt and fructose (Sunett[®] Technical Brochure) (Table 2.3).

There may also be a cost advantage in combining acesulfame-K with

Table 2.3 Sensory properties of intense sweeteners

	Sweetness intensity	Sweetness quality	Synergistic with:
Sucrose	1	Clean sweetness; no after-taste	Saccharin, aspartame, cyclamates, stevioside
Sodium saccharin	200–700	Sweet; bitter, metallic after-taste	Cyclamates, aspartame, sucrose, isomalt, stevioside, NHDC
Sodium cyclamate	30–140	'Chemical' sweet; no after-taste	Saccharin, sucrose, aspartame, acesulfame-K
Aspartame	160–200	Clean sweetness; sweet after-taste	Saccharin, cyclamates, sucrose, glucose, acesulfame-K, isomalt, stevioside
Acesulfame-K	200	Sweet; slight bitter after-taste	Aspartame, cyclamate
Stevioside	300	Slow onset; lingering, liquorice, bitter after-taste	Sucrose, glucose, fructose, aspartame, glycyrrhizin
Thaumatin	1500–2500	Slow onset; lingering, liquorice after-taste	Saccharin, acesulfame-K, stevioside
NHDC	1500–2000	Delayed onset, cooling, menthol-like taste	Most sweeteners
Sucralose	400–800	Sweet taste close to sucrose, very slight delay and lingering of sweetness	
RTI-001	58	Sucrose-like taste, no after-taste	

Sources: von Rymon Lipinski (1986, 1987); Seltzman *et al.* (1985); Anon (1988); Gelardi (1987); Bakal (1983, 1987); Lindley (1983); Ripper *et al.* (1986); Sunett® Technical Brochure; Higginbotham (1983, 1986); Anon (1987b); Crosby and Wingard (1981).

thaumatin, since this mixture reputedly provides an equivalent taste in some products to that of aspartame, a more costly sweetener (Anon, 1987b).

2.6.3 Health and safety

No adverse reactions in the body to the consumption of acesulfame-K have been found despite extensive safety studies. The Joint FAO/WHO Expert Committee on Food Additives (JECFA) allocated an ADI of 0–9 mg kg^{-1} of body weight in 1983, having found that acesulfame-K was neither mutagenic nor carcinogenic, and with no other toxicological problems. It is not metabolised in the body, is excreted rapidly and completely, and thus has no calorific value and is suitable for diabetics. It is also considered to be non-cariogenic, since the acute oral toxicity of acesulfame-K use is extremely low (von Rymon Lipinski, 1986).

2.6.4 Regulations

Acesulfame-K was first cleared by the Food Additives and Contaminants Committee (FACC) for use in the UK, with effect from 1983. It has been accepted for food use by WHO/FAO and by the Scientific Committee for Foods of the EEC with an ADI of $0-9\,mg\,kg^{-1}$ of body weight. More recently, the FDA granted an ADI of up to $15\,mg\,kg^{-1}$ of body weight. This rate of addition to foods and beverages should be sufficient to safely replace all of the sugar consumed in the diet for most people. In addition to the UK and US, acesulfame-K has been approved in many other countries, including Germany, Russia, Australia, South Africa, Cyprus, Belgium, Denmark and Egypt. Its approval has also been granted for use in toothpaste by Bulgaria and the USSR, although no official approval is required by many countries for its use in oral-hygiene products because of its demonstrated safety (von Rymon Lipinski, 1986).

2.6.5 Applications

Because of its good solubility and stability in aqueous media, acesulfame-K is particularly suitable for sweetening soft drinks. At a given concentration, acesulfame-K imparts a slightly higher sweetness in acid foods and beverages than in neutral ones (Luck, 1981). Sweetness intensity is not reduced in hot drinks, relative to those at room temperature, to the same extent as occurs with other sweeteners (Sunett® Technical Brochure). It can be added to liquid concentrates, and there is no hydrolytic decomposition of stock solutions at the usual pH range of drinks, i.e. at pH 3 and above, over a period of months. Also, because of its heat stability, aceslfame-K can be processed in spray towers or for production of instant beverage powders. As a single sweetener, $400-700\,mg\,l^{-1}$ produces a medium sweetness in drinks. Blending with another sweetener in soft drinks should be arranged so that each contributes 50% of the sweetness to obtain the best taste profile. Acesulfame-K is compatible with all sugars, both chemically and sensorily, and a particularly pleasant taste is achieved upon mixing with other high-intensity sweeteners, especially aspartame and cyclamate (Sunett® Technical Brochure).

Acesulfame-K can readily be used in the production of table-top formulations, as solutions or spray-dried granular or powder preparations, because of its solubility in water and heat stability. Solutions should be adjusted to a pH range of 5.5–6.0 with appropriate buffer systems. Compression into tablets, however, requires the addition of a disintegrant such as low-viscosity carboxymethyl cellulose or polyvinyl pyrrolidone. Effervescent tablets can be produced with the addition of sodium hydrogen carbonate as the carbon dioxide donor and tartaric acid as the acid medium, and benefit from the addition of small amounts of cold-water-soluble gelatin. Effervescent tablets have a good shelf-life stored in a dry place. Acesulfame-K can also be mixed with inert substances, or with citrates, tartrates, lactose and/or sugar alcohols to produce tablet-top powders. Mixing with pure cellulose provides calorie-free dustings that are visually attractive on some products.

Acesulfame-K can be combined with pectins and other gelling agents providing bulk in the production of low-calorie preserves, or with sorbitol in sugar-free jams and marmalades. However, such products are more susceptible to spoilage by microorganisms than sugar-containing preserves, and should either be pasteurised or include 0.05–0.1% potassium sorbate as a preservative where permitted. Acesulfame-K is best added as an aqueous solution, thus aiding even dispersion, in the range of 500–2500 mg kg^{-1} of final product.

Because of its good heat stability, acesulfame-K can be used to replace sugar in confectionery (1000–3000 mg kg^{-1}) and in bakery products (500–2000 mg kg^{-1}) together with polydextrose, disaccharide alcohols, sorbitol or isomalt to provide bulk. Concentrations of 500–600 mg kg^{-1} are recommended in desserts, and 500–3000 mg kg^{-1} in sugar-free chewing gum. In the production of sugar-free ice cream, about 500 mg kg^{-1} may be added to supplement sugar alcohols and to achieve a well-balanced taste. Acesulfame-K does not affect the melting and whipping properties of the ice-cream mix.

Acesulfame-K has applications in other products, too. It can be used to sweeten fruit products, fruit yoghurt, sandwich spreads and pickles. It is also suitable for sweetening pharmaceuticals and oral hygiene products, such as toothpaste and mouthwashes, since it masks the bitter or other unpleasant taste characteristics of other product components (Sunett® Technical Brochure).

2.7 Stevioside

2.7.1 General

Stevioside, sucrose and thaumatin are the only sweeteners extracted and refined from plants without chemical or enzymic modification (Phillips, 1978). Stevioside is a sweet glycoside extracted from the leaves of the plant *Stevia rebaudiana* Bertoni, a variety of chrysanthemum found wild in areas of Paraguay and Brazil. The plant has been successfully cultivated in other countries including Japan, Korea, Taiwan and China.

Although other grades of purity exist, stevioside is commercially available in Japan in three basic forms: crude extract, 50% pure, and 90% pure or higher. The taste profile improves with increasing purity. However, a compound that has even better organoleptic properties, i.e. less after-taste, than 90% stevioside is Rebaudioside A. Rebaudioside A is a constituent of stevioside, that has been isolated and the subject of US and Japanese patent applications (Bakal and O'Brien Nabors, 1986).

2.7.2 Mixtures

As a single sweetener, stevioside produces an unacceptable liquorice-like taste in cola beverages. Its combination with fructose, though, has been successfully used in Japan to produce calorie-reduced (50%) soft drinks (Bakal, 1987). Stevioside is combined with sugar alcohols in sugarless chewing gums, and with sucrose in calorie-reduced sugar cubes. The mixture of stevioside with glycyrrhizin is synergistic and is available commercially from Japanese manufacturers. Synergism has also been found (Table 2.3) with aspartame, cyclamate and acesulfame-K, but not with saccharin (Bakal and O'Brien Nabors, 1986).

2.7.3 Health and safety

The results of standard short-term tests, with rats and silkworms from several laboratories, indicate no significant mutagenic or genotoxic activity for stevioside. Human experience of long and extensive use, particularly in Japan, suggests the safety and lack of toxicity of stevioside (Bakal and O'Brien Nabors, 1986). However, it remains uncertain as to whether stevioside and rebaudioside are degraded in the human bowel to steviol, with associated biological risks.

2.7.4 Regulations

Clearance is not needed in Japan for natural products, and stevioside is a permitted additive there as well as in Brazil and Paraguay. Safety evaluations have not yet been completed in Western countries, however.

2.7.5 Applications

Stevioside is currently used in Japan in sugarless chewing gums, soft drinks, table-top sweeteners, juices and other products. Gums sweetened with stevioside enjoy high consumer acceptance, despite the differences between these products and saccharin and aspartame-sweetened products (Bakal, 1987). Stevioside is also added, in Japan, to products such as pickles, dried seafoods, fish, meat and bean pastes and soy sauce as a flavour modifier and to suppress pungent flavours (Bakal and O'Brien Nabors, 1986).

2.8 Thaumatin

2.8.1 *General*

Thaumatin is a mixture of intensely sweet-tasting proteins extracted from the fruit of a West African plant, *Thaumatococcus daniellii*. The two major sweet-tasting proteins, thaumatin I and II (TI and TII), were isolated by Van der Wel and his group at Unilever in 1972. Thaumatin is marketed in the UK by Tate & Lyle plc as Talin, although the fruit of the plant has been used for centuries by the West Africans as a source of sweetness. It is also sold in Japan. Because of problems with stability, taste profile and compatibility, thaumatin is used primarily as a flavour enhancer, at levels below the sweet-taste threshold.

2.8.2 *Mixtures*

Thaumatin is synergistic with saccharin, and masks saccharin after-taste when used at low levels. Synergism is also found with acesulfame-K and with stevioside (Table 2.3), but not with cyclamate or aspartame (Higginbotham, 1986). A combination of thaumatin and acesulfame-K is said to provide a less costly alternative to aspartame with equivalent taste in some products (Anon, 1987b).

2.8.3 *Health and safety*

Thaumatin is the only natural high-intensity sweetener, and products containing it do not require to be labelled 'artificially sweetened'. It has a low calorific value and is non-cariogenic (Higginbotham, 1986). The report of the Joint FAO/WHO Expert Committee (1987) recorded no mutagenic, teratogenic or allergenic effects of thaumatin, and concluded that the lack of toxicity, together with its ready digestion to normal food components, indicated that its only dietary effect was to make an insignificant contribution to the normal protein intake.

2.8.4 *Regulations*

Thaumatin has been permitted as a natural food in Japan since June 1979. It was awarded GRAS status for use in chewing gum in the USA in October

1984 and, in the UK, was permitted for use in foods, drinks and dietary products, excluding baby foods, by the Sweeteners in Foods Regulations in 1983. The Joint FAO/WHO Expert Committee declared an ADI 'not specified' for thaumatin in 1985 (Joint FAO/WHO, 1987). Approval has also been gained in many countries world wide for use of thaumatin as a sweetener and flavour enhancer, particularly in chewing gum. These include Australia, Belgium, Spain, Switzerland, Mexico and Denmark (Higginbotham, 1986).

2.8.5 Applications

As a sweetener, thaumatin is used in beverages and desserts, but its applications are limited because of its liquorice taste and delayed sweetness (Gelardi, 1987). In practice, therefore, thaumatin is more commonly used as a partial sweetener, mixed with other more rapidly tasting sweeteners (Higginbotham, 1986).

Despite its limitations as a sweetener, thaumatin is a powerful flavour enhancer, and magnifies spearmint, cinnamon, wintergreen and peppermint by up to ten times. This flavour potentiating characteristic, together with the lingering sweet taste, can be beneficially used for products such as toothpaste, mouthwash and chewing gum, and for enhancing the masking flavours in medicines. Thaumatin also boosts the low sweetness of bulk sweeteners added to sugarless gums, without adding calories or cariogenicity (Higginbotham, 1983, 1986).

Thaumatin has been used in Japan since 1979 in a variety of products, where it has been shown to enhance and improve the flavour of coffee and of milk products. It is thus used in coffee-flavoured products, ice-cream, iced milk drinks-on-sticks, and spray-dried milk powders. It also enhances savoury flavours (Higginbotham et al., 1981; Higginbotham, 1986), and combinations of thaumatin with nucleotides, spices and/or other flavours may be used to replace monosodium glutamate, an ingredient of current concern with regard to safety (Anon, 1987b).

2.9 Neohesperidin dihydrochalcone

2.9.1 General

In the late 1950s Horowitz and Gentili discovered that a bitter flavanone, hesperetin, found in the peel of the Seville orange, could be chemically converted by alkaline hydrogenation to a sweet compound, neohesperidin dihydrochalcone (NHDC). A straightforward synthetic route has now been developed as an alternative method of its production. Time-intensity studies show that the taste profile of NHDC is unlike that of sucrose, with a delayed onset of menthol-like sweetness and a lingering after-taste (Crosby and Wingard, 1981).

2.9.2 Mixtures

The mixture of NHDC with saccharin produces synergistic sweetness (Table 2.3) and an improved taste profile in soft drinks (Bakal, 1987). Mixing NHDC with other ingredients, such as cream of tartar, a carbohydrate bulking agent and vanilla flavour reputedly eliminates delayed onset and after-taste, while the addition of gluconates, amino acids or nucleotides improves the sweetness (Higginbotham, 1983).

2.9.3 Health and safety

Safety studies have been conducted with laboratory animals mainly in the US over the past fifteen years. The results have suggested that NHDC is neither toxic, mutagenic, carcinogenic, teratogenic nor cariogenic (Horowitz and Gentili, 1986). These results are possibly unsurprising since flavenoids are common constituents of the diet, and only minute quantities of NHDC would normally be consumed.

2.9.4 Regulations

NHDC is currently permitted as a sweetener in several countries, including Belgium, where it may be added to chewing gum and some beverages (Horowitz and Gentili, 1986). Regulatory clearance has been sought in Israel, America and Spain to develop and manufacture NHDC, since these countries have large resources of citrus materials. However, toxicological data presented to the FDA did not conform to the required guidelines, and GRAS status was not awarded.

2.9.5 Applications

NHDC is incompatible with other flavour components in most food applications, including tea and coffee, and is a poor substitute for sucrose because of the delayed onset of sweetness and the menthol-like taste (Crosby and Wingard, 1981). Up to 25% of the sweetness in soft drinks can be contributed by NHDC before the taste becomes unacceptable. However, it provides an acceptable sweetness profile and flavour-enhancing properties in chewing gum, candies, toothpastes and mouthwashes, since these are products that benefit from long-lasting sweetness. NHDC is said to preserve the flavour and aroma of chewing gum at low levels (Higginbotham, 1983). NHDC may be combined with bulking agents such as sugar alcohols in some applications, particularly in view of its high potency.

Although the sensory properties of NHDC-sweetened products and those of products sweetened with saccharin or aspartame are different, consumer acceptance is high in Japan (Bakal, 1987). Applications of NHDC include sweetening cultured milk products, suppressing salt taste in highly brined traditional Japanese foods, and as a tobacco flavourant (Higginbotham, 1983).

NHDC has the propensity for reducing bitterness, as well as providing sweetness, and is thus suitable for inclusion in bitter-tasting drugs and in grapefruit juice (Horowitz and Gentili, 1986).

2.10 Sucralose

2.10.1 *General*

Sucralose is a non-caloric, high-intensity sweetener currently under development. It is derived by the selective chlorination of sucrose at the molecular 4, 1′, and 6′ positions by a patented process developed by Tate & Lyle, London. The sweetener is being jointly developed by Tate and Lyle Speciality Sweeteners in the UK and McNeil Speciality Products Company, a subsidiary of Johnson and Johnson, in the US (Anon, 1988).

2.10.2 *Health and safety*

A comprehensive range of safety evaluation studies has shown that sucralose is not toxic, carcinogenic, teratogenic or mutagenic (Daniel and Das, 1980; Anon, 1988). It is also non-caloric and is not conducive to the formation of dental caries (Higginbotham, 1983).

2.10.3 *Regulations*

Approval of sucralose as a food additive is being sought in a number of countries including the UK, Canada and the US.

2.10.4 *Applications*

Successful applications of sucralose in a wide range of products was reported by Miller, of McNeil Speciality Products Company (Anon, 1988). These include still and carbonated beverages, dry milk products, frozen foods, chewing gum, baked products, fruit spreads and syrups.

2.11 RTI-001

2.11.1 *General*

A new peptide sweetener has been developed at the Research Triangle Institute (RTI), in the US, that is 58 times sweeter than sucrose, has a pleasant sucrose-like taste, and is more stable in aqueous media than aspartame (Seltzman *et al.*, 1985). A US patent (Patent no. 4,714,619) was applied for by Seltzman *et al.*, of the RTI, in December 1987. The authors recommend that, on account of its taste, stability and non-toxicity, this sweet compound should have further consideration as a sucrose substitute. Seltzman further predicts its availability in 5–10 years (Anon, 1986).

2.11.2 *Health and safety*

The compound RTI-001 was tested at the Research Triangle Institute by means of the Ames assay and mouse toxicity studies, and was shown to be non-mutagenic and non-toxic (Seltzman *et al.*, 1985).

2.11.3 *Applications*

The stability of RTI-001 in aqueous and acid media (Seltzman *et al.*, 1985), with twice the shelf-life of aspartame (Anon, 1986), demonstrates its potential for use in soft drinks.

References

Andres, C. (1987) Twin prong aspartame expansion. *Food Processing* (February), 40–45.

Anon (undated) *Sunett® Technical Brochure.* Sunett® Information Service, Hoechst AG, Frankfurt.

Anon (1984) Report of the Scientific Committee for Food on Sweeteners, EC Document III/1316/84/CS/EDUL/27, review.

Anon (1986a) Sugar and substitutes. *Sugar Journal* (September), 19–22.

Anon (1986b) Sweetener competition benefits processors, *Food Engineering Int'l* (March), 23–26.

Anon (1987a) Directly compressible maltitol powder and process for preparing it. French Patent Application IV 285.

Anon (1987b) High intensity sweeteners. *Food Review* (August/September), 35–37.

Anon (1988) Future ingredients – Focus of OVIFT meeting. *Food Technology* (January), 60–64.

Anon (1989a) *Food Facts, Chemistry and Industry* (3 July), 421.

Anon (1989b) News Briefs, *Chemistry and Industry* (18 September), 575.

Anon (1989c) Ingredient news – Low energy sweetening agent. *Food Australia* (May), 730.

Anon (1989d) Ringing the changes, *Food Processing* (January), 2–28.

Bakal, A. (1983) Functionality of combined sweeteners in several food applications, *Chemistry and Industry* (19 September), 700–708.

Bakal, A. (1987) Saccharin functionality and safety. *Food Technology* (January), 117–118.

Bakal, A. J. and O'Brien Nabors, L. (1986) Stevioside. In *Alternative Sweeteners* (eds O'Brien Nabors, L. and Gelardi, R. C.), Marcel Dekker, New York.

Booy, C. J. (1987) Lactitol 'A new food ingredient'. *Bulletin of the International Dairy Federation*, **212**, 62–68.

Chief Medical Officer's Committee on Medical Aspects of Food Policy (COMA) (1984) *Diet and Cardiovascular Disease*. HMSO, London.

Chief Medical Officer's Committee on Medical Aspects of Food Policy (COMA) (1989) *Dietary Sugars and Human Disease*. HMSO, London.

Code of Federal Regulations, Title 21 (April 1989), Part 189.135, p. 499.

Crosby, G. A. and Wingard, R. E. Jr. (1981) A survey of less common sweeteners. In *Developments in Sweeteners – 1* (eds Hough, C. A. M., Parker, K. J. and Vlitos, A. J.), Elsevier Applied Science, London.

Daniel, J. W. and Das, I. (1980) Biological and metabolic studies with the novel sweetener 4,1′,6′-trichloro-,4,1′,6′-trideoxygalactosucrose (abstract), Society of Toxicology, Nineteenth Annual Meeting, Washington, DC, 9–13 March 1980.

den Uyl, C. H. (1987) Technical and commercial aspects of the use of lactitol in foods as a reduced-calorie bulk sweetener. In *Developments in Sweeteners – 3* (ed. Grenby, T. H.), Elsevier Applied Science, London.

Dodson, A. G. and Pepper, T. (1985) Confectionery technology and the pros and cons of using non-sucrose sweeteners. *Food Chemistry* **16**, 271–280.

Dwivedi, B. K. (1986) Polyalcohols: Sorbitol, mannitol, maltitol and hydrogenated starch hydrolysates. In *Alternative Sweeteners* (eds O'Brien Nabors, L. and Gelardi, R. C.), Marcel Dekker, New York.

Fabry, I. (1987) Malbit and its applications in the food industry. In *Developments in Sweeteners – 3* (ed. Grenby, T. H.), Elsevier Applied Science, London.

Federal Register, Title 21 (October 1969), Vol. 34, No. 202, p. 17063.

Garriga, M. M. and Metcalfe, D. D. (1988) *Annals of Allergy* **61**, (6, Part 2), 63–69.

Gelardi, R. C. (1987) The multiple sweetener approach and new sweeteners on the horizon. *Food Technology* (January), 123–124.

Grenby, T. H. (1983) Nutritive sucrose substitutes and dental health. In *Developments in Sweeteners – 2* (eds Grenby, T. H., Parker, K. J. and Lindley, M. G.), Elsevier Applied Science, London.

Havenaar, R. (1987) Dental advantages of some bulk sweeteners in laboratory animal trials. In *Developments in Sweeteners – 3* (ed. Grenby, T. H.), Elsevier Applied Science, London.

Higginbotham, J. D. (1979) Protein sweeteners. In *Developments in Sweeteners – 1* (eds. Hough, C. A. M., Parker, K. J. and Vlitos, A. J.), Elsevier Applied Science, London.

Higginbotham, J. D. (1983) Recent developments in non-nutritive sweeteners. In *Developments in Sweeteners – 2* (eds Grenby, T. H., Parker, K. J. and Lindley, M. G.), Elsevier Applied Science, London.

Higginbotham, J. D. (1986) Talin protein (Thaumatin). In *Alternative Sweeteners* (eds O'Brien Nabors, L. and Gelardi, R. C.), Marcel Dekker, New York.

Higginbotham, J. D., Lindley, M. and Stephens, P. (1981) Flavour potentiating properties of Talin sweetener (thaumatin). In *The Quality of Foods and Beverages – Vol. 1* (eds Charalambous, G. and Inglett, G.), Academic Press, London.

Horowitz, R. M. and Gentili, B. (1986) Dihydrochalcone sweeteners from citrus flavanones. In *Alternative Sweeteners* (eds O'Brien Nabors, L. and Gelardi, R. C.), Marcel Dekker, New York.

Hough, C. A. M. (1979) Sweet polyhyric alcohols. In *Developments in Sweeteners – 1* (eds Hough, C. A. M., Parker, K. J. and Vlitos, A. J.), Elsevier Applied Science, London.

Joint FAO/WHO (1987) Toxicological evaluation of certain food additives and contaminants, Twenty-ninth Meeting of the Joint FAO/WHO Expert Committee on Food Additives, Geneva, June 1985. Cambridge University Press, Cambridge.

Kasperson, R. W. and Primack, N. (1986) Cyclamate. In *Alternative Sweeteners* (eds O'Brien Nabors, L. and Gelardi, R. C.), Marcel Dekker, New York.

Lindley, M. G. (1983) Non-nutritive sweeteners in food systems. In *Developments in Sweeteners – 2* (eds Grenby, T. H., Parker, K. J. and Lindley, M. G.), Elsevier Applied Science, London.

Lindley, M. G. (1991) Sweeteners in markets, marketing, and in product development. In *Handbook of Sweeteners* (eds Marie, S. and Piggott, J. R.), Blackie & Son, Glasgow.

Linke, H. A. B. (1987) Sweeteners and dental health: The influence of sugar substitutes on oral microorganisms. In *Developments in Sweeteners – 3* (ed. Grenby, T. H.), Elsevier Applied Science, London.

Luck, E. (1981) Acesulfame-K. *Ann. Fals. Exp. Chim.* 74, 293.

Mackay, D. A. M. (1987) The food manufacturer's view of sugar substitutes. In *Developments in sweeteners – 3* (ed. Grenby, T. H.), Elsevier Applied Science, London.

Mazur, R. H. and Ripper, A. (1979) Peptide-based sweeteners. In *Developments in Sweeteners – 1* (eds Hough, C. A. M., Parker, K. J. and Vlitos, A. J.), Elsevier Applied Science, London.

Miller, W. T. (1987) The legacy of cyclamate. *Food Technology* (January), 116.

NACNE (1983) A discussion paper on proposals for nutritional guidelines for health education in Britain, prepared for the National Advisory Committee of Nutrition Education by an ad hoc working party under the chairmanship of Professor W. P. T. James, NACNE, London.

Pepper, T. and Olinger, P. M. (1988) Xylitol in sugar-free confections, *Food Technology* (October), 98–106.

Phillips, K. C. (1987) Stevia: Steps in developing a new sweetener. In *Developments in Sweeteners – 3* (ed. Grenby, T. H.), Elsevier Applied Science, London.

Price, J. M. *et al.*, (1969) *Science* **167**, 1131.

Ripper, A., Homler, B. E. and Miller, G. A. (1986) Aspartame. In *Alternative Sweeteners* (eds O'Brien Nabors, L. and Gelardi, R. C.), Marcel Dekker, New York.

Rockstrom, E. (1980) Lycasin hydrogenated hydrolysates. In *Carbohydrate Sweeteners in Foods and Nutrition* (eds Koivistoinen, P. and Hyvonen, L.), Academic Press, London.

Seltzman, H. H., Hsieh, Y. A., Cook, C. E., Hughes, T. J. and Hendren, R. W. (1985) Peptide sweeteners. Abstract, presented to a meeting of the American Chemical Society, 29 April 1985, Miami.

Sicard, P. J. (1982) Hydrogenated glucose syrups, sorbitol, mannitol and xylitol. In *Nutritive Sweeteners* (eds Birch, G. G. and Parker, K. J.), Elsevier Applied Science, London.

Sicard, P. J. and Leroy, P. (1983) Mannitol, Sorbitol, and Lycasin: Properties and food applications. In *Developments in Sweeteners – 2* (eds Grenby, T. H., Parker, K. J. and Lindley, M. G.), Elsevier Applied Science, London.

Stegink, L. D. (1987) Aspartame: Review of the safety issues. *Food Technology* (January), 119–121.

Strater, P. J. (1986) Palatinit – Technological and processing characteristics. In *Alternative Sweeteners* (eds O'Brien Nabors, L. and Gelardi, R. C.), Marcel Dekker, New York.

Sweeteners in Foods (Scotland) Amendment (1988) of Regulation 1983 (SI 1983, No. 1497) under the Food and Drugs (Scotland) Act 1956. HMSO, SI 1988, No. 2084.

United States Code Annotated, Title 21 (Foods and Drugs), Sections 1–800, pocket park 1989, 75–76.

von Hertzen, G. and Lindqvist, C. (1980) Comparative evaluation of carbohydrate sweeteners. In *Carbohydrate Sweeteners in Foods and Nutrition* (eds Koivistoinen, P. and Hyvonen, L.), Academic Press, London.

von Rymon Lipinski, G. W. (1982) Acesulfame-K – new and interesting prospects. *Soft Drinks* (October), 434–435.

von Rymon Lipinski, G. W. (1983) Intense sweeteners. *Food Marketing and Technology* (June), 8–10.

von Rymon Lipinski, G. W. (1985) The new intense sweetener Acesulfame-K. *Food Chemistry* **16**, 259–269.

von Rymon Lipinski, G. W. (1986) Acesulfame-K. In *Alternative Sweeteners* (eds O'Brien Nabors, L. and Gelardi, R. C.), Macel Dekker, New York.

von Rymon Lipinski, G. W. (1987) Intense sweeteners, *Food Marketing & Technology* (June), 8–11.

Walter, G. J. and Mitchell, M. L. (1986) Saccharin. In *Alternative Sweeteners* (eds O'Brien Nabors, L. and Gelardi, R. C.), Marcel Dekker, New York.

Ziesenitz, S. C. and Siebert, G. (1987) The metabolism and utilisation of polyols and other bulk sweeteners compared with sugar. In *Developments in Sweeteners – 3* (ed. Grenby, T. H.), Elsevier Applied Science, London.

3 Flavourings

R. J. GORDON

3.1 Introduction

The sole purpose of the use of food flavourings is to make a finished product more appealing. There are five main reasons for using flavourings:

1. The process involved in the production of the food product may necessitate the addition of flavours. For example, because of the loss of flavour due to heating while baking or the lack of the roast note in microwave cooking.
2. The availability of a natural flavouring ingredient may be unreliable thus necessitating the use of flavours. Poor weather conditions can drastically reduce crop yields for example.
3. Economic factors may restrict the use of natural materials. This could be due to the cost of the natural material itself as in the case of maple syrup.
4. The form of the natural material may not permit it to be used in the finished product. Ginger roots cannot be used as such to produce a beverage but must be extracted for use by a soft drink bottler.
5. The potency of the natural material may be such that it cannot be used practically in the finished product. Flavouring confectionery or baked goods is not practical using fruits or fruit juices alone.

In order to meet the needs of the processed food industry, the following three catagories of flavours are available:

Full or complete flavours. These are flavourings which, on their own, provide the full flavour impact of the named fruit or product (excluding the basic tastes of sweet, acidic, salty or bitter. Example: strawberry flavour in candy, fruit flavours in chewing gum.

Flavour enhancers or modifiers. These are flavours which, when added to a food product, enhance the overall flavour impact of that product. For example, flavours with acetaldehyde and orange essences will enhance the flavour of an orange beverage.

Flavour extenders. These are flavours which themselves may not be indicative of a material but, when added to that material, may allow a reduction of its level use in a finished product without any noticeable change in taste. For example, a flavour which allows for a reduction of cocoa powder in a finished product from 3% to 2% without any noticeable difference in taste. In Tables 3.1 and 3.2 the following are described:

1. The suitability of the various types of flavours (by physical form) for different end use applications.
2. For each application, a brief description of the function of the flavour (i.e. full flavour, enhancer/modifier or extender) along with a general range of dosage range.

Table 3.1 Flavour application suitability by physical form. A – Applicable; B – Use in consultation with supplier; N – Not recommended

Application	Flavour physical form				
	Liquid				
		Water soluble			
	Oil or oil soluble	Ethanol based	Propylene glycol or glycerol based	Emulsion	Powder
Animal food					
Feed	NR	B	B	NR	A
Regular pet food	A	B	A	B	B
Semi-moist pet food	NR	B	A	NR	A
Baked goods					
Bars	B	B	A	B	A
Biscuits	A	B	A	B	A
Cakes	A	A	A	B	A
Cake mixes	NR	NR	NR	NR	A
Crackers	A	B	A	NR	A
Wafers	B	B	A	B	B
Beverages					
(1) Alcoholic:					
Beer/Beer coolers	NR	A	A	A	NR
Liqueurs/cordials	NR	A	A	A	NR
Spirits	NR	A	A	A	NR
Wine/Wine					
coolers	NR	A	A	A	NR
(2) Carbonated:					
With juice	NR	A	A	A	NR
Without juice	NR	A	A	A	NR
(3) Non-carbonated:					
HTST	NR	A	A	A	NR
Juice	B	A	A	A	NR
Pasteurised	NR	A	A	A	NR
Powder mixes	NR	NR	NR	NR	A
Squash	NR	A	A	A	NR
Syrups	NR	A	A	A	NR
UHT	NR	A	A	A	NR

Table 3.1 *Continued*

Application	Oil or oil soluble	Ethanol based	Propylene glycol or glycerol based	Emulsion	Powder
			Flavour physical form		
		Liquid			
		Water soluble			
Confectionery					
Chewing gum	A	NR	A	NR	A
Chocolate	A	NR	B	NR	A
Compound coating	A	NR	B	NR	A
Comprimate	B	A	A	NR	A
Cream centres	A	A	A	B	B
Fondant	A	A	A	NR	B
Fudge	A	A	A	NR	A
Gum drops	A	A	A	NR	NR
Hard-boiled candy	A	A	A	NR	B
Jellies	A	A	A	NR	NR
Jubes	A	A	A	NR	NR
Liquorice	A	B	B	NR	NR
Marshmallow	NR	B	B	NR	A
Nougat:					
White	NR	B	A	NR	A
Brown	A	A	A	NR	A
Tablets	B	A	A	NR	A
Toffee:					
Whipped	A	A	A	NR	A
Caramel	A	A	A	NR	A
Culinary					
Gravy:					
Powder mix	NR	NR	NR	NR	A
Ready to use	B	B	A	B	B
Meat products:					
Sausages, hams, etc.	B	B	B	B	B
Sauces:					
Powder mix	NR	NR	NR	NR	A
Ready to use	B	B	B	B	B
Soups:					
Powder mixes	NR	NR	NR	NR	A
Ready to eat	NR	B	A	B	B
Surimi	NR	B	A	A	B
Dairy					
Cheese:					
Cottage	B	A	A	B	B
Processed	B	A	A	NR	B
Curd	B	A	A	NR	B
Margarine	A	A	A	NR	NR
Milk drink:					
Ready to drink	B	A	A	A	NR
Instant powder	NR	NR	NR	NR	A
Yoghurt	NR	A	A	A	NR
Yoghurt drink:					
Ready to drink	B	A	A	A	NR
Instant powder	NR	NR	NR	NR	A

Table 3.1 *Continued*

Application	Oil or oil soluble	Flavour physical form			
		Liquid			
		Water soluble			
		Ethanol based	Propylene glycol or glycerol based	Emulsion	Powder
Desserts					
Custard:					
Ready to eat	NR	A	A	B	A
Mix	NR	NR	NR	NR	A
Flan	B	A	A	B	B
Gelatin desserts:					
Ready to eat	NR	A	A	NR	A
Mix	NR	NR	NR	NR	A
Mousse	NR	B	B	B	A
Pudding Mix					
Cooked	NR	NR	NR	NR	A
Instant	NR	NR	NR	NR	A
Desserts (Frozen)					
Frozen pops	NR	A	A	A	NR
Ice-cream	NR	B	A	A	NR
Ice-milk	NR	B	A	A	NR
Sherbet	NR	B	A	A	NR
Sorbet	NR	B	A	A	NR
Yoghurt (frozen)	NR	B	A	A	NR
Mouthcare					
Lip balm	B	B	B	NR	NR
Mouthwash	B	A	A	NR	NR
Mouthspray	NR	A	A	NR	NR
Toothpaste	A	A	A	NR	B
Pharmaceuticals					
Cough syrups	NR	A	A	B	NR
Medicated chewing gum	A	B	B	NR	A
Throat lozenges	A	A	A	NR	B
Vitamins – Chewable	B	B	B	NR	A

Table 3.2 General guide to flavour function and dosage, by application. XXX – Most often; XX – Occasionally; X – Rarely

Application	Flavour used as			
	Full flavour	Enhancer or modifier	Extender	Guide dosage range
Animal food				
Feed	XXX	X	XX	0.1 to 0.4%
Regular pet food	XX	XXX	XXX	0.1 to 0.25%

Table 3.2 *Continued*

Application	Flavour used as			Guide dosage range
	Full flavour	Enhancer or modifier	Extender	
Animal food continued				
Semi-moist pet food	XX	XXX	XXX	0.1 to 0.25%
Baked goods[1]				
Bars	XXX	X	XX	0.1 to 0.25%
Biscuits	XXX	X	XX	0.1 to 0.25%
Cakes	XXX	X	XX	0.1 to 0.25%
Cake mixes	XXX	X	XX	0.15 to 0.8% (in cake mix)
Crackers	XX	XX	XXX	0.5 to 2%
Wafers	XXX	X	X	0.1 to 0.25%
Beverages[2]				
1. Alcoholic:				
Beer/beer coolers	XXX	XX	X	Concentrated flavours and extracts 0.05 to 0.2%; emulsions 0.2 to 0.6%; natural flavours 0.5 to 2%
Liqueurs (Cordials)	XXX	XX	X	Concentrated flavours and extracts 0.05 to 0.2%; natural flavours 0.5 to 2%
Spirits	XX	XX	X	Concentrated flavours and extracts 0.05 to 0.2%; natural flavours 0.5 to 2%
Wine/wine coolers	XXX	XX	X	Concentrated flavours and extracts 0.05 to 0.2%; emulsions 0.2 to 0.6%; natural flavours 0.5 to 2%
2. Carbonated:				
With juice	XX	XX	XXX	Concentrated flavours and extracts 0.05 to 0.2%; emulsions and dilute natural flavours 0.25 to 1%
Without juice	XXX	X	X	Concentrated flavours and extracts 0.05 to 0.2%; emulsions and dilute natural flavours 0.25 to 1%
3. Non-carbonated:				
HTST[3]	XXX	X	XX	Concentrated flavours and extracts 0.05 to 0.2%; emulsions and dilute natural flavours 0.25 to 1%

Table 3.2 *Continued*

Application	Full flavour	Enhancer or modifier	Extender	Guide dosage range
		Flavour used as		
Beverages continued				
Juice	X	XXX	XXX	Concentrated flavours and extracts 0.03 to 0.1%; emulsions and dilute natural flavours 0.15 to 0.75%
Pasteurised	XXX	X	XXX	Concentrated flavours and extracts 0.5 to 0.2%; emulsions and dilute natural flavours 0.25 to 1%
Powder mixes	XXX	X	XX	Concentrated flavours 0.03 to 0.15% in finished beverage; dilute natural flavours 0.25 to 1% in finished beverage
Squash	XX	XX	XXX	Concentrated flavours and extracts 0.05 to 0.2%; emulsions and dilute natural flavours 0.25 to 1%
Syrup	XXX	X	X	Concentrated flavours and extracts 0.05 to 0.2%; emulsions and dilute natural flavours 0.25 to 1%
UHT	XX	XX	XXX	Concentrated flavours and extracts 0.05 to 0.2%; emulsions and dilute natural flavours 0.25 to 1%
Confectionery				
Chewing gum[4]	XXX	X	X	0.5 to 1%
Chocolate[5]	X	XXX	XXX	0.05 to 0.15% (powdered flavours may go up to 0.25%)
Compound coating[5]	X	XXX	XXX	0.05 to 0.15% (powdered flavours may go up to 0.25%)
Comprimates[6]	XXX	X	X	0.05 to 0.15% (powdered flavours may go up to 1%)
Cream centres	XXX	X	X	0.05 to 0.15%
Fondant	XXX	X	X	0.05 to 0.15%
Gum drops	XXX	X	X	0.1 to 0.2%
Hard-boiled candy[7]	XXX	X	X	0.05 to 0.15%

Table 3.2 *Continued*

Application	Flavour used as			Guide dosage range
	Full flavour	Enhancer or modifier	Extender	
Confectionery continued				
Jellies	XXX	X	XX	Dosage depends upon gelling agent: Pectin 0.06 to 0.12% Agar-Agar 0.08–0.15% Gelatin 0.1 to 0.2%
Jubes	XXX	X	XX	0.05 to 0.15%
Liquorice[8]	XXX	X	X	0.5 to 1%
Marshmallow	XXX	X	X	0.1 to 0.2%
Nougat[9]	X	XXX	XX	0.12 to 0.25%
Tablets[10]	XXX	X	X	0.1 to 0.25%; powdered flavours may go up to 1%
Toffee:				
Caramel	XXX	XX	X	0.15 to 0.25%
Whipped[11]	XXX	X	X	0.15 to 0.25%
Culinary[12]				
Gravy:				
Powder mix	XXX	XX	X	Concentrated flavours 0.05 to 0.2%; dilute natural flavours 0.25 to 1%[13]
Ready to use	XXX	XX	XX	Concentrated flavours 0.05 to 0.2%; dilute natural flavours 0.25 to 1%
Meat products	X	XXX	XX	Concentrated flavours 0.05 to 0.2%; dilute natural flavours 0.25 to 1%
Sauces	XXX	X		Concentrated flavours 0.05 to 0.2%; dilute natural flavours 0.25 to 1%
Soups:				
Powder mix	XXX	XX	XX	Concentrated flavours 0.05 to 0.2%; dilute natural flavours 0.25 to 1%
Ready to eat	XXX	XX	XX	Concentrated flavours 0.05 to 0.2%; dilute natural flavours 0.25 to 1%[13]
Surimi	XXX	X	X	Concentrated flavours 0.15 to 0.3%; dilute flavours 0.5 to 3%

Table 3.2 *Continued*

Application	Flavour used as			Guide dosage range
	Full flavour	Enhancer or modifier	Extender	
Dairy[14]				
Cheese:				
Cottage	XX	X	XXX	Concentrated flavour 0.05 to 0.15%; dilute natural flavour 0.25 to 1%
Processed	XX	X	XXX	Concentrated flavour 0.05 to 0.15%; dilute natural flavour 0.25 to 1%
Curd	XX	X	XXX	Concentrated flavour 0.05 to 0.15%; dilute natural flavour 0.25 to 1%
Margarine	XXX	X	X	Concentrated flavour 0.05 to 0.15%; dilute natural flavour 0.25 to 1%
Milk drink				
Ready to drink	XXX	X	XX	Concentrated flavour 0.5 to 0.15%; dilute natural flavour 0.25 to 1%
Instant powder	XXX	X	X	Concentrated flavour 0.05 to 0.15%; dilute natural flavour 0.25 to 1%[13]
Yoghurt	XXX	X	XX	Concentrated flavour 0.05 to 0.15%; dilute natural flavour 0.25 to 1%
Yoghurt drink:				
Ready to drink	XXX	X	XX	Concentrated flavour 0.05 to 0.15%; dilute natural flavour 0.25 to 1%
Instant powder	XXX	X	X	Concentrated flavour 0.05 to 0.15%; dilute natural flavour 0.25 to 1%[13]
Desserts				
Custard:				
Ready to eat	XXX	XX	X	0.05 to 0.2%
Mix	XXX	XX	X	0.05 to 0.2%[13]
Flan	XXX	X	X	0.05 to 0.15%
Gelatin Desserts:				
Ready to eat	XXX	X	X	0.05 to 0.2%
Powder mix	XXX	X	X	0.05 to 0.2%[13]

Table 3.2 *Continued*

Application	Full flavour	Enhancer or modifier	Extender	Guide dosage range
		Flavour used as		
Desserts continued				
Mousse:				
Ready to eat	XXX	X	XX	0.05 to 0.2%
Powder mix	XXX	X	XX	0.05 to 0.2%
Pudding mix:				
Cooked	XXX	X	XX	0.05 to 0.2%[13]
Instant	XXX	X	XX	0.05 to 0.2%[13]
Desserts (Frozen)				
Frozen pops	XXX	X	X	Concentrated flavours 0.05–0.15%; dilute natural flavours and emulsions 0.5–1%
Ice-cream	XXX	X	XX	Concentrated flavours 0.08–0.2%; dilute natural flavours and emulsions 0.5–2%
Ice-milk	XXX	X	X	Concentrated flavours 0.05 to 0.15%; dilute natural flavours and emulsions 0.5 to 1.5%
Sherbet	XXX	X	X	Concentrated flavours 0.05 to 0.15%; dilute natural flavours and emulsions 0.5 to 1.5%
Sorbet	XXX	X	X	Concentrated flavours 0.05 to 0.15%; dilute natural flavours and emulsions 0.5 to 1.5%
Yoghurt (frozen)	XXX	X	XX	Concentrated flavours 0.08–0.2%; dilute natural flavours and emulsion 0.5–2%
Mouthcare				
Lip balm	XXX	X	X	0.75 to 1.5
Mouthwash	XXX	X	X	0.07% to 0.2%[15]
Mouthspray	XXX	X	X	2 to 4%[15]
Toothpaste	XXX	X	X	0.75 to 3%[16]
Pharmaceuticals				
Cough syrups	XXX	X	X	Concentrated flavours 0.1 to 0.4%; dilute flavours 1 to 2%
Medicated chewing gum	XXX	X	X	0.25 to 1%
Throat lozenges	XXX	X	X	0.1 to 0.2%
Vitamins – Chewable	XXX	X	X	1 to 3%[17]

Table 3.2 *Continued*

Application	Flavour used as			Guide dosage range
	Full flavour	Enhancer or modifier	Extender	
Snacks				
Coated	XXX	XX	X	0.25 to 1%[18]
Extruded:				
Post-extrusion	XXX	X	X	Concentrated flavour 0.05 to 0.2%; dilute flavour 0.4 to 1%; dusting powder 5 to 12%
Pre-extrusion	XXX	X	XX	0.1 to 1%

Notes:

[1]Flavours must be heat stable. Dosage is affected by duration and temperature of baking.

[2]Some flavours, particularly emulsions, may cause cloudiness in the finished product which can be desirable for some products.

[3]Beverages processed by pasteurisation UHT or HTST require flavours that are process stable. Flavour supplier should be notified of the process.

[4]In chewing gum, there could be a problem of compatibility between gum base and flavour.

[5]With chocolate or compound coating, the flavour could affect the viscosity of the finished product. In chocolate and compound coating, powder flavours may be used at up to 0.25%.

[6]Use of citrus oils in comprimates may result in shorter product stability. Powdered flavour dosage may go up to 1%.

[7]Flavours with a dosage of 0.7% are not suitable for hard-boiled candies.

[8]High impact flavours are required for liquorice.

[9]Fruit juices, pastes, essential oils, dilute flavours and some solvent may adversely affect foam structure.

[10]Liquid flavours are not recommended for effervescent tablets.

[11]High alcohol content flavours or flavours with a dosage over 0.7% are not suitable for whipped toffee.

[12]Products processed by hot packing/heat treatment or intended for microwave use require flavours suitable for this process.

[13]In ready-to-eat product.

[14]Flavours are used in most dairy applications to provide non-dairy flavour notes, for example strawberry or mango flavour for yoghurt, etc.

[15]Mint flavours or mint combinations are the most popular flavour types.

[16]Topnote flavours dosage in toothpaste will be 0.15–0.3%.

[17]Powdered flavours are recommended for this application.

[18]Liquid flavours are commonly dispersed in sugar-gum solution for coating cereal. Powdered flavours may be used, particularly for citrus and mint flavours.

3.2 Glossary

Absolute   The alcoholic extract of a concrete, used to separate the essential oil from the waxy material portion of the concrete.

Agglomerate   A powdered flavour produced by spraying a liquid flavouring substance onto the surface of a powdered carrier which is being

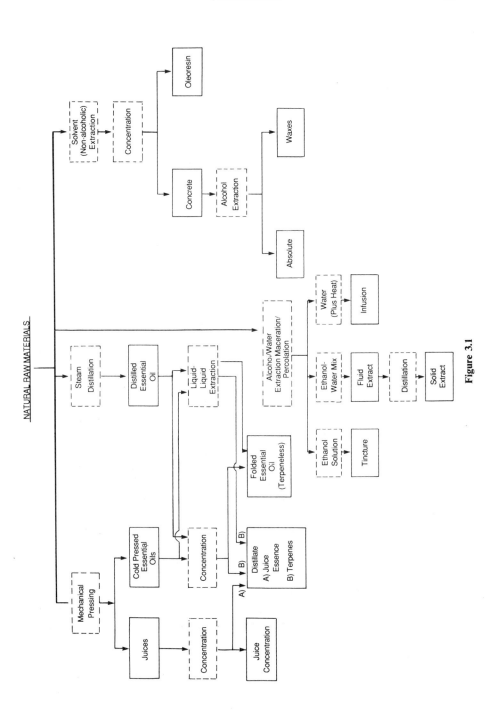

Figure 3.1

sprayed into the chamber of the agglomerater. This is a continuous process.

Adsorbate A powdered flavour which is made by coating a liquid flavouring onto the surface of a powdered carrier such as corn starch, salt or maltodextrin. This process is carried on batch-wise.

Balsam The natural exudate from a tree or plant.

Citrus oil The essential oil obtained from the peel of a citrus fruit such as lemon, orange or lime.

Concrete This is the semi-solid mixture of essential oil and fatty, waxy material and is obtained by the volatile solvent extraction of flowers or plants followed by solvent removal.

Distillate The volatile material recovered by condensing the vapours of an extract or press-cake of fruits which is heated to its boiling point in a still. This distillation may be under ambient or vacuum conditions.

Distillation The physical process of heating a product to the boiling point in a still and collecting the vapours by condensing them through cooling. Distillation may be done under ambient or vacuum conditions and the desired product may be the distillate, the residue left in the still or both.

Distilled oil An essential oil obtained by the distillation of the portion of a botanical material, e.g. peel, leaves, stem, containing the essential oil.

Emulsion A liquid flavouring which is a mixture of essential oils, gum acacia and/or other gums or modified food starch in water. This is a means of utilising the flavouring properties of an essential oil in a water-based food product. An emulsion may contain added colour.

Essence oil The oil recovered from the distillate that is obtained during the concentration of citrus juices. These products are usually quite fragrant.

Essential oil The active flavouring principles of certain botanicals such as roots, stems, leaves and buds of spices and herbs, seeds, flowers, citrus fruit skins and barks of certain trees. The oil is found in small oil sacs which are distributed throughout the plant structure concerned and normally, the oil bears close resemblance to the parent plant.

Extract-fluid The product obtained by the extraction of certain botanicals with a liquid solvent, usually ethanol and water. The botanicals do not contain oils and the flavouring principles are soluble in water or the solvent used.

Extract-solid A viscous or semi-solid material obtained by first extracting the botanical material with a water–ethanol solvent and then removing the solvent almost completely. This is a liquid extract which has been concentrated.

Flavouring A preparation made with or without solvents or carrier, used exclusively to impart flavour with the exception of sweet, salty, acidic or bitter tastes. It is not intended to be consumed as such or simply diluted for consumption. It can be either in liquid or powdered form.

Flavour enhancer   A substance which has little or no odour by itself at the dosage used and the primary purpose of which is to increase the flavour effect of certain food components well beyond any flavour effect contributed by the substance itself, e.g. monosodium glutamate.

Fluid extract   See Extract-fluid.

Fold   Refers to the ratio of the quantity of starting material to finished product and indicates the strength of the finished product, e.g. a five-fold oil is one in which 100 kg of natural oil has been concentrated to 20 kg. The concentration can be done by distillation or other evaporative techniques (most common) or by other, more sophisticated processes such as liquid–liquid extraction, followed by distillation. In the case of vanilla extracts, the strength is regulated in many countries including Canada and the United States.

Infusion   A liquid extract prepared by using a hot water or hot water–ethanol solution, which is refluxed over the raw material.

Isolate   An aromatic compound consisting of one ingredient isolated from a natural raw material such as menthol from peppermint oil or citral from lemongrass oil.

Juice concentrate   The concentration of a fruit juice by the removal of water by evaporative techniques such as falling film evaporator or distillation.

Juice essence concentrate   The product that is made by the concentration of the distillate obtained from the preparation of a juice concentrate. The distillate may be concentrated 100 to 300 fold and is normally quite fragrant.

Liquid flavouring   A flavouring in liquid form which may or may not contain a solvent.

Oleoresin   The semi-solid material obtained by first extracting a spice with a volatile solvent (e.g. hexane) and then totally removing the solvent by distillation. It contains both the essential oil and the resinous material.

Powdered flavouring   A flavour which is produced either by spray-drying, adsorption, agglomeration, dry-blending or other such process.

Sesquiterpene   A family of chemicals which is based upon three isoprene units. As with terpenes, these chemical compounds occur naturally in essential oils, particularly citrus oils. They have little flavour, poor water solubility and react readily with oxygen to produce off-aroma and off-flavour notes. Sesquiterpenes have a higher boiling point than mono-terpenes.

Solid extract   See Extract-solid.

Spice oil   The essential oil obtained from a spice such as nutmeg, clove or cinnamon.

Spray-dried flavour   A powdered flavour which is made by first preparing an emulsion of the flavouring to encapsulate the flavouring with gum acacia or modified food starch followed by spraying this emulsion into a spray drying unit where the water is rapidly evaporated and the dry product collected. This is a continuous process.

Steam distillation The passing of steam through a bed of plant material with the essential oil being carried over with the water vapour by co-distillation and then collected after passing through a condenser. The water and oil are then physically separated, the oil dried and clarified.

Terpenes A family of chemicals based upon two isoprene units:

These chemicals make up a major portion of many essential oils, particularly citrus oils. These compounds have very little flavour, poor water solubility and react readily with oxygen to produce off-aroma and off-flavour notes.

Terpeneless oil This is an essential oil from which a good portion of the terpenes have been removed. This is done by distillation or by liquid–liquid extraction (terpenes have poor solubility while the desirable oxygenated compounds are more soluble in ethanol solutions) followed by distillation.

Tincture A solution in ethanol of the flavour principles of certain botanicals obtained by maceration or percolation.

Acknowledgements

The author wishes to thank Messrs G. Sinki, G. Novel, A. Geissler, T. Tjan and particularly Ms. Elizabeth Ho for their assistance in preparing this chapter.

Further reading

Arctander, S. (1960) *Perfume and Flavour Materials of Natural Origin.* S. Arctander, Elizabeth, N. J.

Fenaroli, G. (1975) *Handbook of Flavour Ingredients,* CRC Press, Cleveland, Ohio.

Givaudan (1989) *A Short Introduction to Flavours by Givaudan.* Givaudan Dubendorf Ltd, Switzerland.

Guenther, E. (1948) *The Essential Oils.* Van Nostrand, New York.

Merory, J. (1968) *Food Flavourings – Composition Manufacture and Use.* AVI Publishing, Westport, Conn.

Ockerman, H. (1978) *Source Book For Food Scientists.* AVI Publishing, Westport, Conn.

Tressler, D. and Sultan, W. (1975) *Food Products Formulary.* AVI Publishing, Westport, Conn.

4 Colours

P. RAYNER

Food and drink have been coloured for centuries. Until the discovery of dye synthesis in 1856, natural extracts from animal, vegetable and mineral origins were used. An attractively coloured foodstuff stimulates the appetite and enhances the enjoyment. It is also a readily demonstrated fact that colour is extremely important in the perception of flavour. If the colour and flavour of a food are not correctly associated, the taster is more likely to identify the food by its colour rather than its actual flavour.

The pigments present in most fresh fruit and vegetables are vivid and brilliant, making these foods attractive to eat but when they are processed their protective environment in the plant cell is disturbed or destroyed. The natural pigments are then subjected to adverse physical and chemical conditions which can cause their partial degradation and colour change or fade. In the UK and North America about 75% of food is now being processed in some way before it reaches the consumer. Manufacturers must replace lost colour if the accepted attractive appearance is to be restored. It is also desirable to provide colour in products that would otherwise be colourless or have little colour, for example soft drinks, confectionery, instant desserts and ice cream.

Colour is also added to supplement the natural appearance of a food and to ensure batch to batch uniformity where raw materials have varying colour intensities.

Over the last century the colours added to foods have been evaluated for their technical suitability and toxicological harmlessness based on the knowledge available at the time. This has reduced the range of colours available for food use and most countries now have permitted lists of colours, together with restrictions on use with certain products. Although there is some level of harmonisation, most countries still have national legislation so it is essential to check the status of a colour before using it in a food product.

This chapter documents the major colours, both synthetic and natural, that are permitted in various countries.

4.1 Synthetic colours

The synthetic colours available are normally very pure chemicals with standardised colour strengths (Tables 4.1 and 4.2). The water-soluble dyes dissolve in water and certain polyhydric compounds such as propylene glycol. Their colouring power is directly proportional to their chemical dye content. The colours are available as powders, pastes, granules and solutions. Blends are also available and some companies offer pre-weighed colours or blends in the forms above, and also in jelly or fat-based sticks for convenience of use. The general stability of the colours is given in the tables. Water-soluble colours are used in a wide range of food products and offer, in general, good stability, consistency and economy in use. They are normally compatible with each other in blends.

Many of the colours are also available as the aluminium lake of the dye. The aluminium lakes are produced by reacting solutions of the water-soluble colours with freshly prepared alumina (aluminium hydroxide). The lakes are essentially insoluble in water and organic solvents. They are produced as extremely fine powders with various dye contents from about 40 to 10%. In general, they have enhanced light stability over the soluble dye and can be used to colour dry powder products, snack products, chewing gum (where there is less staining of the mouth), sugar confectionery (especially as suspensions with titanium dioxide for coating sweets or tablets) and compressed tablets.

4.1.1 Summary of typical usages for synthetic colours

Tartrazine (yellow) General purpose. Powdered desserts, confectionery, ice-cream and dairy products, soft drinks, pickles, sauces, fish and baked products.

Yellow 2G (bright yellow) General purpose colour.

Quinoline yellow (bright, greenish yellow) General purpose colour. Soft drinks, desserts, confectionery, ice-cream and dairy products.

Sunset yellow FCF (yellow orange) General purpose colour. Soft drinks (but can precipitate out if calcium ions are present), ice-cream, confectionery, canned foods, baked goods, and desserts.

Orange RN (bright reddish orange) General purpose colour.

Orange G (bright orange) General purpose colour.

Carmoisine (bluish red) Confectionery, soft drinks, ice cream, desserts, canned fruit.

Ponceau 4R (bright red) Soft drinks, confectionery, jellies, canned goods, fish and as lake to colour cheese rind and coated confections.

Amaranth (bluish red) Canned goods, soft drinks, jams, ice-cream and powdered desserts.

Table 4.1 Properties of synthetic colours (CI = Colour Index (British); FD&C = Food, Drug and Cosmetic Act – USA)

Colour	CI 1971	EEC No.	FD&C No.	Fastness						Solubility			
				Light	Heat (°C)	Alkali	Fruit acid	Benzoic acid	Sulphur dioxide	Water	Glycerol	Ethanol	Propylene glycol
Yellows/Oranges													
Tartrazine[1]	19140	E102	Yellow 5	VG	VG to 105°	G	VG	VG	VG	10	7	SS	2
Yellow 2G	18965	—	—	VG	VG to 205°	reddens VG	VG	VG	VG	8	MS	SS	INS
Quinoline yellow[1]	47005	E104	—	G	G to 105°	P	VG	P	VG	14	SS	SS	SS
Sunset yellow (Or/Yel) FCF[1]	18965	E110	Yellow 6	VG	VG to 205°	F reddens	VG	VG	F	10	4	SS	1
Orange RN	15970	—	—	G	VG to 105°	F reddens	G	G	F	4	5	SS	SS
Orange G	16230	—	—	VG	F to 205°	F reddens	VG	G	F	6	5	SS	INS
Reds													
Carmoisine[1] (Azorubine)	14710	E122	—	G	G to 105°	F	VG	G	F	4	2	SS	1
Ponceau 4R[1]	16255	E124	—	G	VG to 105°	G	G	G	F	12	SS	SS	4
Amaranth[1]	16185	E123	—	G	VG to 105°	F bluer	VG	G	F	7	1	SS	SS
Red 2G	18050	128	—	VG	G to 205°	VG	G	G	G	6	1	INS	3
Erythrosine[1]	45430	E127	Red 3	F	G to 105°	F bluer	P	P	G	6	3	1	16
Allura red (FD&C red No. 40)	16035	—	Red 40	G	VG to 105°	F	VG	G	F	12	2	SS	2
Blues													
Indigo carmine[1]	73015	E132	Blue 2	F	F to 105°	P	F	P	P	1	SS	SS	SS
Patent blue V[1]	42051	E131	—	G	G	F	P	P	F	4	3	SS	20
Brillant blue FCF[1]	42090	133	Blue 1	G	VG to 105°	G	VG	VG	VG	20	5	SS	20

Table 4.1 *Continued*

Colour	CI 1971	EEC No.	FD&C No.	Fastness						Solubility			
				Light	Heat (°C)	Alkali	Fruit acid	Benzoic acid	Sulphur dioxide	Water	Glycerol	Ethanol	Propylene glycol
Greens													
Green S (Brilliant green BS)	44090	E142	—	P	VG to 205°	P	G	G	G	5	1	SS	2
Fast green FCF	42053	—	Green 3	F	F	P	F	F	G	20	20	20	20
Browns													
Brown FK	—	154	—	F	G to 105°	VG	G	G	G	20	5	SS	INS
Chocolate brown FB	—	—	—	F	VG to 205°	VG	VG	VG	G	15	5	INS	INS
Chocolate brown HT	20285	155	—	G	VG to 205°	VG	VG	VG	F	20	5	INS	15
Blacks													
Brilliant black BN	28440	E151	—	G	F	P	G	G	P	5	SS	SS	INS
Black 7984	27755	—	—	F	F	—	F	F	F	5	SS	INS	INS

Fastness
VG – very good
G – good
F – fair
P – poor
Heat fastness: to 105°C or 205°C
[1]Lake pigments available.

Solubility (e.g. 10 = 10 g (100 ml)$^{-1}$)
INS – insoluble
SS – slightly soluble
MS – moderately soluble
S – soluble

N.B. The stability data should only be taken as a guideline since products may be subjected to more than one of the phenomena detailed. The stability of the colour is also dependent on the concentration used. Solubility data should only be taken as a guideline since the solubility is dependent on the dye content, physical state of the colour and the manufacturing methods.

Table 4.2 Data on synthetic colours

Colour	Hue	Chemical type	Chemical name	Empirical formula
Tartrazine	Yellow	Monoazo	Trisodium 5-hydroxy-1-(4-sulphonatophenyl)-4-(4-sulphonatophenylazo)pyrazole-3-carboxylate.	$C_{16}H_9N_4O_9S_2Na_3$
Yellow 2G	Bright	Monoazo	1-(2,5-dichloro-4-sulphophenyl)-3-methyl-5-pyrazolone	$C_{16}H_{10}N_4S_2O_7Cl_2Na_2$
Quinoline yellow	Bright greenish yellow	Quinophthalone	Sodium salts of a mixture of disulphonates (principally), monosulphonates and tri-sulphonates of 2-(2 quinolyl)indan-1,3-dione	$C_{18}H_9NO_8S_2Na_2$
Sunset yellow FCF	Yellow Orange	Monoazo	Disodium 6-hydroxy-5-(4-sulphonatophenylazo)-naphthalene-2-sulphonate	$C_{16}H_{10}N_2O_7S_2Na_2$
Orange RN	Bright reddish orange	Monoazo	Sodium 6-hydroxy-5-(phenylazo)naphthalene-2-sulphonate	$C_{16}H_{11}N_2O_4NaS$
Orange G	Bright orange	Monoazo	Disodium 6-hydroxy-5-(phenylazo)-naphthalene-2,4-disulphonate	$C_{16}H_{10}N_2O_7Na_2$
Carmoisine	Bluish red	Monoazo	Disodium 4-hydroxy-3-(4-sulphonato-1-naphthylazo) naphthalene-1-sulphonate	$C_{20}H_{12}N_2O_7S_2Na_2$
Ponceau 4R	Bright red	Monoazo	Trisodium 7-hydroxy-8(4-sulphonato-1-naphthylazo) naphthalene-1,3-disulphonate.	$C_{20}H_{11}N_2O_{10}S_3Na_3$
Amaranth	Bluish red	Monoazo	Trisodium 3-hydroxy-4-(4-sulphonato-1-naphthylazo) naphthalene-2,7-disulphonate	$C_{20}H_{11}N_2O_{10}S_3Na_3$
Red 2G	Bluish red	Monoazo	Disodium 5-acetamido-4-hydroxy-3-phenylazonaphthalene-2,7-disulphonate	$C_{18}H_{13}N_3O_8S_2Na_2$
Erythrosine	Bright bluish red	Xanthene	Disodium 2-(2,4,5,7-tetraiodo-3-oxido-6-oxoxanthen-9-yl)benzoate	$C_{20}H_6O_5I_4Na_2$
Allura Red	Red	Monoazo	Disodium 2-hydroxy-1-(2-methoxy-5-methyl-4-sulphonato-phenylazo)naphthalene-6-sulphonate	$C_{18}H_{14}N_2O_8S_2Na_2$

Table 4.2 *Continued*

Colour	Hue	Chemical type	Chemical name	Empirical formula
Indigo carmine	Blue	Indigoid	Disodium 3,3'-dioxo-2-2'-bi-indolylidene-5,5'-disulphonate, disodium 3,3'-dioxo-2,2'-bi-indolylidene-5,7'-disulphonate	$C_{16}H_8N_2O_8S_2Na_2$
Patent blue V	Bright blue	Triarylmethane	Sodium compound of [4,4-diethylaminophenyl)-5-hydroxy-2,4 disulphonphenyl-methlidene]2,5-cyclohexadien-1-ylidene] diethyl-ammonium hydroxide inner salt	$C_{27}H_{31}N_2O_7S_2Na$
Brillant blue FCF	Bright greenish blue	Triarylmethane	Disodium [4-(N-ethyl-3-sulphonatobenzylaminophenyl]-[4-(N-ethyl-3-sulphonatobenzyl-iminio)cyclohexa-2,5-dienylidene]toluene-2-sulphonate.	$C_{37}H_{34}N_2O_9S_3Na_2$
Green S	Bluish green	Triarylmethane	Sodium 5-[4-dimethylamino-(4-dimethyliminiocyclo-hexa-2,5-dienylidene)benzyl]-6-hydroxy-7-sulphonatonaphthalene-2-sulphonate	$C_{27}H_{25}N_2O_7S_2Na$
Fast green FCF	Bluish green	Triarylmethane	N-ethyl-N-(4-[(4-(ethyl)((3-sulphophenyl)methyl)amino)phenyl) (4-Hydroxy-2-sulphophenyl)methylene)-2,5-cyclohexadien-1-ylidene)-3-sulphobenzene methanaminium hydroxide, inner salt disodium salt	$C_{35}H_{30}N_2O_{10}S_3Na_2$
Brown FK	Yellowish brown		Essentially a mixture of the sodium salts of 4,4'-(4,6-diamino-m-phenylenebisazo) dibenzenesulphonic acid and the sodium salt of 4(4,6-diamino-m-tolylazo)benzenesulphonic acid	
Brown FB	Brown		Monoazo dye produced by coupling diazotised naphthionic acid with a mixture of (CI 75240 and CI 75660)maclurin and morin	
Brown HT	Reddish brown	Diazo	Disodium 4,4'-(2,4-dihydroxy-5-hydroxymethyl-1,3-phenylene bisazo) di(naphthalene-1-sulphonate)	$C_{27}H_{18}N_4S_2O_9Na_2$
Brillant black BN	Violet	Diazo	Tetra-sodium 4-acetamido-5-hydroxy-6[7-sulphonato-4-(4-sulphonatophenylazo)-1-naphthylazo]naphthalene-1,7-disulphonate	$C_{28}H_{17}N_5O_{14}S_4Na_4$
Black 7984	Violet	Diazo	Tetra-sodium 3-amino-5-hydroxy-6 [7-sulphonato-4-(4-sulphonato phenylazo)-1-naphthalazo]naphthalene-2,7-disulphonate	$C_{26}H_{15}O_{13}S_4Na_4$

Red 2G (bluish red)   Meat products, sugar confectionery and jams.

Erythrosine (bright bluish red)   Because it forms erythrosinic acid in solutions of pH 3 to 4, and *this* is only slightly soluble, it is the only red colour used with cherries which does not colour other fruits. Used for glacé and maraschino cherries, meat products, confectionery and canned goods.

Allura red (red)   General purpose colour.

Indigo carmine (blue)   Confectionery.

Patent blue V (bright blue)   General purpose colour. Confectionery drinks and icing.

Brilliant blue FCF (bright greenish blue)   General purpose colour. Confectionery drinks and icing.

Green S (bluish green)   General purpose colour. Often blended with yellow to produce leaf green hues.

Fast green FCF (bluish green)   General purpose colour, often blended to produce various shades.

Brown FK (yellowish brown)   Used for colouring fish in brine without precipitations.

Chocolate brown FB (brown)   Mainly for baked cereal products, sugar confectionery and desserts.

Chocolate brown HT (reddish brown)   General purpose colour. Baked products, vinegar and confectionery.

Brilliant black BN (violet)   General purpose colour generally used in blends, also fish roe products and confectionery.

Black 7894 (violet)   General purpose colour.

4.2 Natural colours

With the discovery of synthetic food colours the demand for natural colours diminished, since the natural extracts lacked consistency of shade; often lacked light and heat stability; were subject to supply variation; and often proved more difficult to use since they would react with the food product and each other. For various reasons the demand for natural colours has dramatically increased in the last few years and the technology of extraction and use has concomitantly developed. Many natural food colours are regularly consumed as part of the average diet, for example lutein and chlorophylls in green vegetables, and anthocyanins, the principle red colours in many fruits.

It is convenient to divide the plant pigments currently used into several groups to discuss their general properties. When considering a natural colour for food products it is necessary to check both the legislative status of the extract for the country concerned, and the particular food product. It is also necessary to understand which specific pigments will be compatible with the ingredients, process conditions, packaging, and desired shelf-life, and with each other when blended to achieve the desired shades.

In most instances, the natural extracts commercially available are not pure pigments but are mixtures of pigments found in nature. The actual shade will depend on the specific source, the method of extraction and purification, and any subsequent blending carried out to achieve a fairly standard usable colour form.

Most of the principal natural colour extracts are available in liquid form or as spray-dried powders. As there are few standard strengths, and it is often difficult to determine the actual pigment contents, the suppliers should be consulted for guidance to usage levels of particular extracts. In many instances the extracted pigments are not water soluble and it is necessary to solubilise them for food applications. The oil-soluble pigments can be solubilised by blending them with suitable permitted emulsifiers, by emulsifying them with permitted stabilisers or by spray drying an emulsion so that the colour is finely dispersed in the food or drink when used.

4.2.1 Summary of typical usages for natural colours

Anthocyanins (blue red shades)   Soft drinks, alcoholic drinks, sugar confectionery (not marzipan or fondant), preserves, fruit toppings and sauces, pickles, dry mixes, canned and frozen foods, dairy products.

Annatto (orange shades)   Oil soluble bixin: dairy and fat based products, processed cheeses, butter, margarine, creams, desserts, baked and snack foods. Water soluble norbixin: sugar and flour confectionery, cheese, smoked fish, ice-cream and dairy products, desserts, custard powders, cereal products and bread crumbs.

β-Carotene (yellow to orange)   Butter, margarine, fats, oils, processed cheeses. Water dispersible forms: soft drinks, fruit juices, sugar and flour confectionery, ice-cream, yoghurts, desserts, cheese, soups and canned products.

Apocarotenal (orange to orange red)   Cheese, sauces, spreads, oils, fats, ice-cream, cake mixes, cake toppings, snack foods and soft drinks.

Canthaxanthin (orange red to red)   Sugar confectionery, sauces, soups, meat and fish dishes, ice-cream, biscuits, bread crumb and salad dressings.

Paprika (orange to red)   Meat products, snack seasoning, soups, relishes, salad dressings, processed cheeses, sugar confectionery, fruit sauces and toppings.

Saffron (yellow)   Baked goods, rice dishes, soups, meat dishes and sugar confectionery.

Crocin (yellow)   Smoked white fish, dairy products, sugar and flour confectionery, jams and preserves, rice and pasta dishes.

Lutein (yellow)   Salad dressings, ice cream, dairy products, sugar and flour confectionery, especially marzipan; and soft drinks.

Beet red (bluish red)   Frozen and short shelf-life foods, ice cream, flavoured milks, yoghurts, dry mix desserts, jelly crystals and slabs.

Cochineal (orange)   Soft drinks and alcoholic drinks.

Cochineal carmine (bluish red)   Soft drinks, sugar and flour confectionery, flavoured milks, desserts, sauces, canned and frozen products, pickles and relishes, preserves and soups.

Sandalwood (orange to orange red)   Fish processing, alcoholic drinks, seafood dressings, bread crumbs, snack seasonings and meat products.

Alkannet (red)   Sugar confectionery, ice cream and alcoholic drinks.

Chlorophyll (olive green)   Sugar confectionery, soups, sauces, fruit products, dairy products, pickles and relishes, jams and preserves, pet foods and drinks.

Copper chlorophyll (green)   Flour and sugar confectionery, soups, sauces, pickles, relishes, fruit products, ice cream, yoghurt, jelly, desserts, dry mix desserts, sauces and soups, soft drinks.

Caramel   Alcoholic and soft drinks, sugar and flour confectionery, soups, sauces, desserts, dairy products, ice-cream, dry mixes, pickles and relishes.

Malt extract   Alcoholic and soft drinks, sugar and flour confectionery, soups, sauces, desserts, dairy products, ice-cream, dry mixes, pickles and relishes.

Turmeric (bright yellow)   Ice cream, yoghurt, frozen products, pickles and relishes, flour and some sugar confectionery, dry mixes and yellow fats.

Riboflavin (yellow)   Cereal products, sugar coated confectionery, sorbet and ice-cream.

Vegetable carbon black (black)   Sugar confectionery and shading colour.

Orchil (red)   Soft drinks, alcoholic drinks, sugar confectionery.

Safflower (yellow)   Soft drinks and alcoholic drinks.

4.3 Flavonoids

Flavonoids are a large group of compounds widely distributed throughout nature. They include quercetin, present in onion skins, and anthocyanins, the major commercially used group.

4.3.1 *Anthocyanins*

One of the most important and widely distributed groups of water-soluble natural colours, anthocyanins (EEC No. E163) are responsible for the attractive red, purple and blue colours of many flowers, fruits and vegetables. Over 200 individual anthocyanins have been identified, of which 20 have been shown to be naturally present in black grapes, the major commercial source of anthocyanin pigment for food coloration.

Chemically, anthocyanins are the glucosides of anthocyanidins and are based on the 2-phenyl benzopyrylium (flavylium) structure. The six most common anthocyanins are: Pelargonidin (Pg), Cyanidin (Cy), Delphinidin (Dp), Petunidin (Pt), Peonidin (Pn), Malvidin (Mv).

Anthocyanin pigments are obtained by extraction with acidified water or alcohol followed by concentration under vacuum and/or reverse osmosis. The extracts can be spray or vacuum dried to give a powder. Extracts from different sources exhibit different colour shades and levels of stability in use (Table 4.4).

Table 4.3 Sources of major anthocyanins

Source	Major anthocyanins present
Grape skins (*Vitis vinifera*)	Cy, Du, Dp, Pt, Mv, monoglycosides
Grape lees (*Vitis labrusca*)	free and acetylated
Cranberry (*Vaccinium macrocarpon*)	Cy and Pu monoglycosides
Roselle calyces (*Hibiscus sabdariffa*)	Cy and Dp mono and biosides
Red cabbage (*Brassica oleracea*)	Cy glycosides
Elderberry (*Sambucus nigra*)	Cy glycosides
Blackcurrant (*Ribes nigrum*)	Cy and Dp mono and diglycosides free and acetylated
Purple corn (*Maiz morado*)	Pg, Cy and Pn monoglycosides

Properties. Anthocyanins are natural indicators (Table 4.4); in acidic media they appear red but as the pH changes they progressively become bluer. They are most stable between pH 2 to 5. Anthocyanins will complex with metal ions resulting in a bluer shade and will also naturally condense with other flavonoids. Although the latter increases their resistance to bleaching by sulphur dioxide, it can result in complexing with proteins and the precipitation of naturally present tannins. This is of special significance if they are used in gelatine-containing products.

Usage is normally 10–40 ppm pure pigment.

Suggested applications: Soft drinks, alcoholic drinks, sugar confectionery, preserves, fruit toppings and sauces, pickles, dry mixes, canned and frozen foods, dairy products.

4.4 Carotenoids

Carotenoids form one of the most important groups of natural pigments responsible for many of the striking yellow, orange and red colours of edible fruits, vegetables, flowers, fungi and some animals (Table 4.5). Over 400 naturally occurring carotenoids have been identified and total annual production in nature is estimated at over 100 million tonnes. The most common are fucoxanthin (in marine algae), lutein (grasses), violaxanthin, β-carotene, lycopene and apocarotenoids. Carotenoids are also found combined with proteins, as carotenoproteins. Astaxanthin and canthaxanthin normally occur as carotenoproteins in nature.

The majority of carotenoids contain 40 carbon atoms and are built up in nature from an ordered fusion of eight C5 isoprene units. Their characteristic properties and striking colours are due to the presence of a system of conjugated C=C bonds in their structure. They can be divided into the following classes:

- Carotenoid hydrocarbons, e.g. β-carotene and lycopene.
- Xanthophylls, oxygen-containing carotenoids, e.g. lutein and violaxanthin.
- Apocarotenoids, which contain fewer than 40 carbon atoms, e.g. bixin and crocin.

Some carotenoids can be converted by the body into Vitamin A during digestion. This is an essential vitamin for healthy growth. The principal pro-Vitamin A is β-carotene.

A few carotenoids have been synthesised commercially as pure pigments and are available in oil- or water-dispersible form. Most carotenoids are oil soluble but only to a limited extent; they are intense pigments and usage levels are typically less than 10 ppm pure pigment. Commercially water-dispersible forms are prepared by adding emulsifiers and/or making emulsions which can be spray dried.

Natural colour carotenoid extracts rarely consist of a single pigment but are usually mixtures. There are at last 300 known carotenoids, each of which is related to the parent compound lycopene, the red pigment of tomatoes.

4.4.1 Specific carotenoid colours

Annatto is the extract from the pericarp of the seed of the annatto tree, Bixa orellana, grown in South America, East Africa and India. The oil-soluble extract, Bixin, is an orange/yellow colour with good light and heat stability. It is susceptible to oxidation which is accelerated by heat and light. The colour is mainly used in dairy and fat-based products such as processed cheese, butter, margarine, creams, desserts, baked and snack foods. It can be blended with turmeric to produce a more yellow shade or paprika to give a redder shade.

Bixin can be hydrolysed by alkali treatment during or after extraction to produce water-soluble Norbixin, having similar colour and stability. Norbixin will precipitate in low pH conditions and can react with metallic salts in water to give a haze. Norbixin is widely used for colouring cheese, smoked fish, flour confectionery, cereal products, breadcrumbs, ice creams

Table 4.4 Properties of anthocyanins

Anthocyanin	Colour range	Light fastness	Heat stability	Stability to oxidation	Stability to pH change	Stability to microbial attack
Grape skin	Red/purple/ blue	G	G	G		F
Grape (Concord)	Red/purple/ blue	G	G	G		F
Cranberry	Pink-red to brown red	G	G	G		F
Hibiscus or roselle	Bright red to brown to dark green at pH 7	FG	F	G	Red in acid Purple in neutral. Blue in alkali.	F
Red cabbage	Bright red to blue	G	G	G		F
Elderberry	Purple-red to blue	FG	G	G		F
Black currant	Purple-red to blue	G	G	G		F
Purple corn	Purple-red to purple	G	VG	G		F

G = good; VG = very good; F = fair; FG = fairly good; P = poor; VS = very soluble; INS = insoluble.

Table 4.5 Properties of carotenoids

Extract	Colour range	Light fastness	Heat stability	Stability to pH change	Stability to oxidation	Stability to microbial attack	Acid resistance	Alkali resistance
Tomato	Orange to red	Good	Good	Good	Fair	Fair	Good	Good
Annatto-Bixin	Orange to peach	Moderate	Good below 125°C	Good	Fair	Good	Poor	Good
Annatto-Norbixin	Orange to peach	Moderate	Good below 125°C	Good but will precipitate at low pH	Fair	Good	Poor	Good
β-Carotene	Yellow to orange	Fair	Good	Good	Fair	Fair	Good	Good

Table 4.4 *Continued*

Acid resis-tance	Alkali resis-tance	Solubility			Comments	Major pigments	Mol. wt
		Water	Vegetable oil	Ethanol			
G	P	VS	INS	VS		Pelargonidin	271.3
G	P	VS	INS	VS		Cyanidin	287.3
G	P	VS	INS	VS		Delphinidin	303.3
G	P	VS	INS	VS	Bluer shade in presence of metal ions, especially tin, iron and aluminium	Petunidin	317.3
G	P	VS	INS	VS		Peonidin	301.2
G	P	VS	INS	VS			
G	P	VS	INS	VS			
G	P	VS	INS	VS		Malvedin	331.3

Table 4.5 *Continued*

Solubility			Comments	Major pigments	EEC No. CI 1971	Mol. wt.	Major sources
Water	Vegetable oil	Ethanol					
INS	S	—	—	Lycopene	E160(d)	536.85	Tomato (*Lycopersicon solanaceae*)
INS	S	S	—	Bixin	E160b 75120	394.5	Seeds of the Bixa Orellana shrub
S	INS	S	Can react with metal ions to form a haze. SO_2 above 100 ppm can cause fade. Binds with proteins with stabilisation effect	Norbixin		380.5	
INS	SS 0.05 g (100 ml)$^{-1}$	SS	Stable to SO_2 Added ascorbic acid reduces colour loss in soft drinks	β-Carotene	E160a 75130	536.85	Carrots (*Daucus carota*) Alfalfa (*Medicago sativa*) Maize (*Zea mays*) Dunaliella algae Synthesised chemical

Table 4.5 *Continued*

Extract	Colour range	Light fastness	Heat stability	Stability to pH change	Stability to oxidation	Stability to microbial attack	Acid resistance	Alkali resistance
Apocarotenal	Orange to orange red	Fair	Fair	Good	Poor	Fair	Good	Good
Canthaxanthin	Orange red to red	Good	Good	Good	Fair	Fair	Good	Good
Paprika	Red to orange	Fair as oleoresin; poor as powder	Good	Good	Poor	Fair	Good	Good
Saffron	Bright yellow	Poor	Good	Fair	Poor	Fair	Good	Fair
Crocin	Bright yellow	Poor	Good	Fair	Poor	Fair	Good	Fair
Tagetes (Lutein)	Egg yellow	Good	Good	Good	Fair	Fair	Good	Good
BETALAINE PIGMENT								
Beet red	Pink to red	Poor	Poor	Good – most stable pH 3.5–5.0	Liquid fair; powder good	Liquid fair; powder good	Good	Poor
QUININOID PIGMENTS								
Cochineal (carminic acid)	Orange to red	Good	Good	Orange in acid; red to violet in alkali	Good	Good	Good	Ppt
Cochineal carmine	Purple red	VG	VG	Precipitates below pH 3; becomes bluer in alkali	VG	VG	Precipitates below pH 3	Stable
Sandalwood	Orange to red	Good	Good	Fair. Becomes redder in alkali	Good	Good	Good	Good
Alkannet	Red to purple red	—	—	Red below pH 6. Blue above.	—	—	—	—
Kermes								
Lac								

Table 4.5 *Continued*

Solubility							
Water	Vegetable oil	Ethanol	Comments	Major pigments	EEC No. CI 1971	Mol. wt.	Major sources
INS	SS better than β-carotene 0.7 − 1.5 g $(100\,ml)^{-1}$	SS		β-apo-8′ carotenal	E160e 40820	416.65	Synthesised chemical Occurs in skin and flesh of oranges
INS	SS 0.005 g $(100\,ml)^{-1}$	SS	—	Canthaxanthin	E161g 40850	564.82	Synthesised chemical Occurs in edible mush-rooms, *Can-tharellus cinna-barina*, also hydra and brine shrimp
INS	S	—	Has a char-acteristic spicy flavour	Capsanthin Capsorubin	E160c	584.85 600.84	Red peppers. (*Capsicum annuum*)
S	INS	S	Bleached by SO_2 Has a char-acteristic flavour	Crocin Crocetin	75100	977.08 328.39	Saffron crocus (*Crocus sativus*)
S	INS	S	Crocin has no flavour	Crocin	75100	977.08	*Gardenia jasminoides* fruit
INS	S	S	Good stability to SO_2	Lutein	E161b	568.0	Aztec marigold. (*Tagetes erecta*) Alfalfa (*Medicago sativa*) also present in green leaves, vegetables and egg yolk
S	INS	S	Limited stability to SO_2	Betanin	E162	568.5	Beetroot (*Beta vulgaris*)
S	INS	S	Reacts with metal ions to form red lakes	Carminic acid	E120 75470	492.4	Cochineal insect (*Coccus cacti*)
INS	INS	INS	Poorly soluble in alkali	Aluminium lake of above			
INS	—	S		Santalin A & B	75550	—	Extract of the heartwood of the sandalwood tree (*Pterocarpus santalinus*)
SS	INS	S		Alkannin	75530	288.29	Roots of *Alkanna tinctoria* and *Anchusa tinctoria*
				Kermesic Acid	75460	—	Extract of the dried female insect of the species Kermoccus ilices
				Laccaic acids	75450	—	Extract from the exudate of the insect *Coccus laceac* and *Laccifera lacca*

Table 4.5 *Continued*

Extract	Colour range	Light fastness	Heat stability	Stability to pH change	Stability to oxidation	Stability to microbial attack	Acid resistance	Alkali resistance
PORPHYRIN PIGMENTS								
Chlorophyll	Olive green	Poor	Poor	Poor	Poor	Poor	Poor	Better than acidic conditions
Copper chlorophyll complexes	Green	Good	Good	Good	Good	Good	Fair	Good
Copper chlorophyllin complexes	Green	Good	Good	Good	Good	Good	Can precipitate	Good
MELANOIDIN PIGMENTS								
Caramel	Red–brown shades	Good	Good	Good – Depends on type	Good	Good	Depends on type	Depends on type
Malt	Brown shades	Good	Good	Good	Good	Fair	Good	Poor
OTHER PIGMENTS								
Turmeric	Bright yellow	Poor	Good	Fair – becomes more orange in alkali	Mod	Fair	Good	Fair
Riboflavin	Greenish yellow	Poor	Good	Poor	Mod	Fair	Good	Poor
Riboflavin 5'-phosphate	Greenish yellow	Poor	Good	Poor	Mod	Fair	Good	Poor
Vegetable carbon black	Black	VG	VG	VG	VG	VG	VG	VG
Orchil	Red	—	—	—	—	—	—	—
Indigotin	Blue	Poor	Good	—	Good	—	Fair	Poor
Safflower or Cathamus	Yellow	Good	Good	Good	Good	Good	Good	Fair
Monascus								

Table 4.5 *Continued*

Solubility				Major	EEC No.		
Water	Vegetable oil	Ethanol	Comments	pigments	CI 1971	Mol. wt.	Major sources
INS	S	S		Chlorophyll a & b	E140 75810	—	Green plants such as alfalfa, grasses and nettles
INS	S	S		Copper chlorophyll a & b	E141	—	Chemical modification of natural chlorophyll
S	INS	S		Sodium copper chlorophyll	E141 75815	—	Chemical modification of of natural chlorophyll and chlorophyllin
S	INS	S					
S	INS	S					
INS	S	S	Bleached by SO_2	Curcumin	E100 75300	368.4	Rhizomes of *Curcuma longa*
SS	SS	INS	Vitamin B2	Riboflavin (Vitamin B2)	E101	376.36	Synthesised chemical but also present in milk and yeast. Can be biosynthesised.
S	SS	INS		Riboflavin-5'-phosphate sodium salt	101a	478.34	One of the physiologically active forms in which Riboflavin exerts its biological functions in an organism
INS	INS	INS		Carbon	E153	—	Carbonised vegetable matter
INS	INS	S	Sulphurated orchil is water soluble	Orcein	E121	—	Extract from Roccella lichens
S	INS	SS		Indigotin	75781	466.37	Leaves of various species of Indigofera
S	INS	S		Safflomin A or Safflor Yellow A	CI 1975 Natural Yellow 5	450.4	Aqueous extract of the flowers of the Dyer's thistle (*Carthamus tinctorius*)
				Carthamin	CI 1975 Natural Red 26 75140	434.40	Aqueous sodium hydroxide extract of *Carthamus tinctorius* flowers
				Monascin (deep yellow)	CI 1975 Natural Red 2	—	*Monascus anka* and *Monascus purpurens* grown on rice
				Ankaflavin		—	
				Rubropunctatin		—	
				Monascorubin (red)		—	

and desserts. It can complex with proteins giving improved light stability and a redder shade.

β-Carotene is available as a nature identical synthesised colour in oil-soluble or special water-dispersible forms. It is also available as carrot extract, as an extract from the Dunaliella algae and is present in alfalfa, maize and other natural products. It is a major pro-Vitamin A carotenoid. It imparts a yellow to orange colour and has wide applications including butter, margarine, fats, oils, cheese, soft drinks, ice cream, yoghurts, desserts, flour and sugar confectionery, jellies, preserves, dressings and meat products.

Apocarotenal occurs in oranges but is normally available as a synthesised colour. It imparts an orange to orange/red colour. Oil- and special water-dispersible forms are available. It is more sensitive to oxidation than *β*-carotene and less stable to light. Uses include cheese, sauces, spreads, oils, fats, ice cream, cake mixes, cake toppings, snack foods and soft drinks. It is frequently blended with *β*-carotene, since it is redder in shade, to achieve a rich orange shade.

Canthaxanthin occurs naturally but is normally available as a synthesised colour. It imparts a red colour. Its solubility is very poor in oil and it is insoluble in water but special water-dispersible forms are available. It can be used in a wide range of foods such as sugar confectionery, sauces, soups, meat and fish products, ice cream, biscuits, breadcrumbs and salad dressings. It blends well with *β*-carotene to give orange shades.

Paprika is the oil-soluble red to orange colour extracted from the sweet red pepper *Capsicum annum*. The extract normally has a spicy flavour which tends to restrict its use to savoury or highly flavoured products. The oleoresin is used in meat products, such as sausages, for flavour and colour. It is also used in seasonings, snack products, soups, relishes, sugar confectionery, cheese, fruit sauces and toppings. Commercially, the colour strength of oleoresin paprika is given in International Colour Units. These units are determined by spectrophotometric absorbance at 460 nm in acetone. The International Colour Unit is the absorbance of the standard dilution times 65 000. This is related to ASTA units (American Spice Trade Association Units) ASTA units × 40 = International Colour Units.

Saffron extract is the yellow-soluble colour extracted from the stigmas and styles of the *Crocus sativus* flowers. It is an expensive colour since about 75 000 flowers are required to produce 1 lb of saffron, yielding about 25 g of pigment. The flowers are hand picked. The extract is normally used as much for its flavour as for its colour. It is heat stable but sensitive to oxidation and sulphur dioxide levels above 50 ppm can cause bleaching. It can be used in baked goods, rice dishes, soups, meat dishes and sugar confectionery.

Crocin extract is basically the same pigment as occurs in saffron but is commercially extracted from the dried fruit of the *Gardenia jasminoides* bush which grows in the Far East. The dried fruit has been used as a tea infusion in the Orient for many decades. The colour extract does not possess the flavour of saffron and is ideal for the coloration of smoked white fish such as cod and haddock where it binds to the flesh. Other applications include dairy products, sugar and flour confectionery, jams and preserves, rice and pasta dishes. In flour confectionery the crocin extract can react with the enzyme β-glucosidase in the presence of proteins to give a green hue. This occurs over a period of 6–8 hours but the enzyme is inactivated during baking.

Gardenia fruits also yield other pigments of the iridoids and flavonoids groups. These extracts can give red, green and blue colours and they are used in Japan and the Far East to colour confectionery, ices, cakes, noodles, beans and other dishes.

Lutein, a xanthophyll, is one of the most commonly consumed carotenoid colours. It is found in all green leaves, green vegetables, eggs and some flowers. Commercially the Aztec Marigold. (*Tagetes erecta*) yields an oil-soluble yellow extract which is mainly lutein. The same pigment can be extracted from alfalfa. The colour has good stability to heat, light and sulphur dioxide and is less sensitive to oxidation than many other carotenoids. The extracts can be used to colour salad dressings, ice creams, dairy products, emulsified fats and other food products containing high levels of fats. They can also be used in soft drinks, flour and sugar confectionery, especially marzipan.

4.4.2 *Betalaine pigments*

These are the water-soluble pigments responsible for the red and yellow shades found in various varieties of beetroots and also pokeberries.

Beetroot is a vegetable that is a temperate crop cultivated in all north European countries. The principle pigments in the juice are betanin (red) and vulgaxanthin (yellow). The beetroots are prepared, washed and then the juice is extracted by crushing. The juice can then be concentrated under vacuum; roller or spray dried or the sugars can be reduced by aerobic fermentation prior to concentration and/or drying. Beetroot juice concentrate provides a pink hue when diluted with water. It has limited stability to heat, light, oxygen and sulphur dioxide, especially in high water activity systems. It is effective in short shelf-life, frozen and low-moisture foods that are not subjected to prolonged heat processing. It is most stable between pH 3.5 to 5.0. Major uses are ice-cream, fruit yoghurts, flavoured milks and dry mixes.

4.4.3 *Quininoid pigments*

These pigments exist in large numbers in nature and are widely distributed in roots, wood, bark and certain insects. The colours range from yellow through brown to red.

The most important quininoid food colours are cochineal and cochineal carmine. *Cochineal extract* consists mainly of carminic acid which exhibits an orange shade in acidic conditions and red in more neutral media above pH 5 and will turn violet and precipitate at pH 7 and above. The extract is obtained from the dried bodies of female insects (*Coccus cacti*) which live on certain cactus plants. In the traditional purification process the aqueous extract is precipitated as the insoluble aluminium lake which is known as cochineal carmine. Carmine is a bright red colour soluble in alkaline media. It is very stable to heat, light and oxidation. In alkaline conditions carmine provides a blue–red shade which becomes progressively less blue as the pH is decreased. Under acidic conditions, below pH 3, the carmine becomes insoluble. Carmine is widely used as a food colour and usage levels are generally in the range 0.04–0.2%. Other quininoid pigments are listed below, although the last three are no longer in common use as colours.

Sandalwood is a traditional spice imparting an orange–red shade. The ground sandalwood is used in fish processing and the extract can be used to colour alcoholic drinks, seafood dressings, breadcrumbs, snacks, food seasonings and meat products.

Alkannet is extracted from the roots of the plants *Alkanna tinctoria* and *Anchusa tinctoria* found in southern Europe. The red colour is sparingly soluble in water but very soluble in organic solvents. Where permitted it can be used in confectionery, ice-cream and alcoholic drinks.

Kermes is a red colour extracted from the dried female insects of the species *Kermococcus ilices*.

Lac is a red colour extracted from the solidified exudation of the insect *Coccus laccae*, it is water soluble. The exudate is shellac.

Madder, the extract of the root of the plant *Rubia tinctorium*, was used as a red food colour.

4.4.4 *Porphyrin and haem pigments*

Porphyrin pigments are common in nature, being the pigments responsible for the green colour of plants – chlorophylls and chlorophyllins. Haem pigments, responsible for the colour of blood, only occur in the diet when blood is an ingredient of meat products.

4.4.5 *Melanoidin pigments*

Melanoidin pigments form the attractive red/brown colours associated with cooked foods. When carbohydrates are heated at high temperatures they caramelise to produce a characteristic flavour and colour of caramel. These pigments are responsible for the colour of caramels, malt extract and toasted carob flour.

Caramel colours are probably the most widely used colours in food and drinks. Some caramel is still manufactured by heating sucrose in an open pan until it froths and caramelises in a traditional way. However, the majority of caramel is manufactured by heating concentrated carbohydrates such as glucose syrups in the presence of accelerators, sometimes under pressure. After the reaction where the larger molecules are initially hydrolysed and polymerised, the mix is rapidly cooled then blended and standardised.

Positively charged caramel is manufactured using ammonia as an accelerator and can reach a colour level of 60 000 EBC (European Brewing Convention) units. These caramels have an isoelectric point of about pH 6.0. The isoelectric point is the pH at which the caramel has no net charge. The major use for electropositive caramels is in the brewing industry where stability in the presence of positively charged fining colloids is required. When a caramel is introduced into a different pH system it can react with other charged particles in the solution and the polymeric colour molecules can flocculate, causing hazing.

Negatively charged caramel is manufactured using ammonium bisulphite as a catalyst and under acidic conditions. The maximum colour value is around 25 000 EBC units. Electronegative caramels exhibit isoelectric points below pH 3 and are widely used in the soft drinks industry where stability at low pH is required.

Spirit caramel is manufactured using sodium hydroxide as a catalyst and the colour value is around 20 000 EBC units. These caramels possess only a weak ionic charge and are stable in high alcoholic products such as rum and whisky.

Liquid caramels can be spray dried to give powders for use in dry food products.

4.4.6 Other natural colours

Turmeric is the fluorescent yellow coloured extract of the rhizome of the Curcuma plant. The major pigment is curcumin and this can be extracted and purified from the turmeric extract. The basic turmeric extract has a spicy flavour and aroma whereas the curcumin pigment has very little flavour and aroma. Tumeric extracts are oil soluble with neligible water solubility; they are soluble in alkalis, alcohols and glacial acetic acid. The pigment is relatively stable to heat but fades rapidly in light in the presence of oxygen.

The turmeric oleoresins are used in savoury products such as pickles, mustard, canned and dry mix foods. The deodorised curcumin extracts are used in frozen dairy products which only have limited exposure to light. They can also be used in wrapped confectionery and the colour appears to be relatively light stable in high-boiled sugar confections.

Riboflavin (Vitamin B_2) is a yellow pigment present in plant and animal cells, such as milk and yeast. Most of the product commercially available is synthesised. The pigment is slightly soluble in water and ethanol, giving a greenish-yellow fluorescent solution. It is soluble in alkali but the solution decomposes rapidly. It is stable under acidic conditions. Riboflavin is used primarily as a source of vitamin but is also used to colour cereal products and sugar coated confectionery and tablets. It has a bitter taste.

The synthesised *riboflavin-5'-phosphate sodium salt* is more soluble in water, less bitter, but is not quite as light stable.

Vegetable carbon black is produced by fully carbonising vegetable material and grinding it to a very fine powder. It is a heat- and light-stable insoluble pigment used in confectionery and also as a shading colour in other food products.

Orchil is the red colour obtained by ammonia extraction of various species of Roccella lichens and Orchella mosses found in the Azores, Canary Islands, Cape of Good Hope and the Cape Verde Islands. The major pigment is orcein. The pigment, which is insoluble in water but dissolves in alcohol, can also be sulphurated to give a water-soluble colour. The colour was used principally for colouring wine, soft drinks and confectionery.

Indigotin or indigo carmine is a water-soluble deep-blue colour considered a permitted natural colour in a few countries and used to colour foodstuffs such as ice cream, desserts and confectionery. Most indigo carmine is now synthesised. Natural indigo occurs as its precursor the glucoside Indican together with Indirubin in the leaves of various species of Indigofera. The leaves are fermented when the Indican is enzymatically hydrolysed to free indoxyl which is then oxidised by air to indigo. Sulphonation of indigo produces its 5,5'-disulphonic acid derivative indigotin. Some natural indigo is still produced in India.

Safflower or carthamus yellow is the water-soluble extract of the dried petals of the Dyer's thistle *Carthamus tinctorius*. The extract imparts a lemon–yellow colour and a mild flavour to various food products. The colour is stable between pH 3 to 9 showing good heat and light stability, especially for soft drinks.

A red pigment can be obtained from the residue of the petals by aqueous sodium hydroxide extraction. The colour is poorly soluble in water but has been used as a food and cosmetic colour. It is mainly used as a textile dye and was formerly used for colouring Goverment red tape.

4.5 Monascus

The best known of the microbial colours are those produced by the Monascus species, especially those of *Monascus anka* and *Monascus purpureus*. The fungus is traditionally grown on rice until the rice becomes coloured by the red mycelia. The whole mass is then dried and powdered for addition to food or consumed. Monascus pigments have been consumed in the Orient for hundreds of years as a food colour for wine, bean curd and other foods or medicines. The colours are stable between pH 2 to 10 and are heat stable.

Recent research, on general culture and substrates, has been directed at optimising yields and producing a range of colours from yellow to red. In general, the pigments are relatively insoluble in water and are oil soluble. However, water solubility can be increased by reacting the pigments with naturally occurring compounds containing amino groups.

4.6 Algae

The algae biliproteins, also called phycobiliproteins, are a group of pigments that occur in the red algae (Rhodophyta), the blue-green algae (Cyanophyta) and the cryptomonad algae (Cryptophyta). The main categories are the red phycoerythrins and the blue phycocyanins. Pigment preparations soluble in water or alcohol can be prepared and potential applications are sugar confectionery, sherbets, frozen confections, ices and chewing gum.

4.7 Inorganic pigments

Titanium dioxide (Table 4.6) can be obtained from the natural mineral ilmenite ($FeTiO_3$) It is an intensely white insoluble pigment with excellent heat and light stability. A major use is in sugar panned confectionery to give an opaque white finish or as a background to other colours.

Iron oxides and hydroxides provide a range of red, yellow and black colours with excellent heat and light stability. They are insoluble pigments and are used mainly in fish pastes and pet foods where they withstand severe heat treatment during retorting and extruding.

Calcium carbonate occurs extensively in nature as chalk, limestone, marble and feldspar. It is an insoluble pigment that can be used to impart an opaque white appearance. Sometimes used in sugar confectionery instead of titanium dioxide.

Ultramarine is a polysulphide of an alkali metal aluminosilicate. It is an insoluble pigment used to impart a blue colour. Its use is now severely restricted. It is sometimes used in pet foods and was used to whiten sugar.

Silver, gold and aluminium are used as surface coloration in the form of finely divided powder or leaf (very thin sheets) for confectionery items and cake decorations.

Ferrous gluconate is permitted in some countries for enhancing the colour of ripe olives.

Table 4.6 Inorganic pigments

Name	Colour	Pigment	EEC No.	CI No.	Major sources	Light/heat fastness	Solubility		
							Water	Acid, Dilute	Alkali, Dilute
Titanium dioxide	White	TiO_2	E171	77891	Natural mineral ilmenite	VG	INS	Slight	INS
Red iron oxide	Red	Anhydrous ferric oxide	E172	77491	Natural minerals	VG	INS		
Black iron oxide	Black	Ferroso ferric oxide	E172	77499					
Yellow iron oxide	Yellow	Hydrated ferric oxide	E172	77492					
Calcium carbonate	White	$CaCO_3$	E170	77220	Limestone, chalk and marble	VG	SS	SS in hydrochloric acid	INS
Ultramarine	Blue	Polysulphide of alkali metal aluminosilicate		77007	Ground lapis lazuli, but can be synthesised	VG	INS	INS	INS
Silver leaf or powder	Silver	Silver metal (Ag)	E174		Metallic silver	VG	INS	S	INS
Gold leaf	Gold	Gold metal (Au)	E175		Metallic gold	VG	INS	INS	INS
Aluminium leaf or powder	Silver	Aluminium metal (Al)	E173		Metallic aluminium	VG	INS	SS	SS

VG = very good; SS = slightly soluble; INS = insoluble; S = soluble

4.7.1 *Summary of typical usages for inorganic pigments*

Titanium dioxide (white)   Sugar-coated confectionery.

Iron oxides   Sugar-coated confectionery, pet foods, meat and fish pastes.

Silver, gold and aluminium   Surface coating of sugar confectionery and cake decorations.

Further reading

Colour Index (3rd edition) 1971, Vol. 3 and *Colour Index* (3rd edition) 1982, Vol. 7. The Society of Dyers and Colourists, Bradford, UK, and the American Association of Textile Chemists and Colourists North Carolina, USA.

Counsell, J. N. (1981) *Natural Colours for Food and Other Uses.* Applied Science Publishers, Barking, Essex.

Dziezak, J. D. (1987) Applications of food colorants. *Food Technology* (April).

Francis, F. J. (1987) Lesser known food colorants. *Food Technology* (April).

Kassner, J. E. (1987) Modern technologies in the manufacture of certified food colors. *Food Technology* (April).

Knewstubb, C. J. and Henry, B. S. (1988) Natural colours – a challenge and an opportunity. *Food Technology International, Europe 1988.* Sterling Publications, London, UK.

Leadbetter, S. (1989) Natural colours: a literative survey. *Food Focus* (July). Leatherhead Food Research Association, UK.

Marmion, D. M. (1984) *Handbook of U.S. Colorants for Food, Drugs and Cosmetics* (2nd edition). Wiley, New York, Chichester, Brisbane, Toronto, Singapore.

Newsome, R. L. (1986) *Food Colors – A Scientific Status Summary by the Institute of Food Technologists Expert Panel on Food Safety and Nutrition.* Institute of Food Technologists, Chicago, USA.

Otterstätter, G. (1988) Food coloring. *Food Marketing and Technology* (June).

Timberlake, C. F. and Henry, B. S. (1986) Plant pigments as natural food colours. *Endeavour,* New Series, **10**, No. 1.

Walford, J. (1980) *Developments in Food Colours 1.* Applied Science Publishers, Barking, Essex, UK.

Walford, J. (1984) *Developments in Food Colours 2.* Elsevier Applied Science, London and New York.

Technical and sales literature are available from the following companies:

BASF Corporation.	Hilton Davis Company
Biocon UK Limited.	Kalsec Inc.
Butterfield Laboratories Limited.	Morton International
Chr. Hansen's Laboratory Inc.	Overseal Foods Limited.
Colorcon Limited	Quest International
D. F. Anstead Limited.	Warner Jenkinson.
F. Hoffmann La Roche and Company Limited.	

5 Preservatives

J. SMITH

5.1 Introduction

The aim of this chapter is to provide the food scientist with practical information which will enable selection of the appropriate antimicrobial preservatives for use in food products. Although the trend in the market place has been towards less additives in foods, attention must be paid by the professional food scientist to the possibility of survival and proliferation of pathogenic microorganisms in foods. Antimicrobial preservatives can still provide a useful role in combination with good manufacturing practice to ensure safe food products. The use of antimicrobial food preservatives also extends the storage life of foods and reduces wastage, costs of processing and disposal problems.

Many substances are inhibitory to microorganisms. Most are unsuitable for use in foods, due to toxicity, undesirable chemical reactions or flavour effects. Consequently, a restricted range of antimicrobial food additives has emerged, through traditional use and approval or through scientific and commercial endeavour, which meets the criteria set by governments throughout the world. There is discrepancy between the various governments over which antimicrobials are acceptable in which foods and at what levels. Information on permitted foods and levels is not contained in this chapter, but may be found within the published legislation of the appropriate government. Methods of analysis are also normally specified within the legislation. General methods of analysis may be found within the Official Methods of Analysis of the Association of Official Analytical Chemists (Williams, 1984).

Two books which provide useful reviews of the topic are *Antimicrobial Food Preservatives* (Lueck, 1980) and *Antimicrobials in Foods* (Branen and Davidson, 1983). These provide a wider range of information on the antimicrobials covered in this chapter and on others not sufficiently accepted to be considered here. For coverage of antimicrobials in addition to the other factors which affect the life and death of microorganisms consult *Microbial Ecology of Foods*, Volume 1 (ICMSF, 1980).

The overview of the January 1989 edition of *Food Technology* was devoted to 'Antimicrobials and their use in foods'. This overview provides an interesting range of broad reviews by Beuchat and Golden (1989),

Wagner and Moberg (1989), Davidson and Parish (1989), Roberts (1989) and Daeschel (1989).

The antimicrobial preservatives are purchased for use in foods in either dried or concentrated form. Safety in handling is an important consideration and while beyond the scope of this chapter, is thoroughly addressed in *Food Additives Handbook* (Lewis, 1989).

Table 5.1 outlines the properties of antimicrobial preservatives which are generally permitted by governments throughout the world. Legislation usually restricts the type of foods in which they may be used and the levels of addition.

Table 5.2 covers the use of other substances which are less commonly considered specifically as antimicrobial preservatives, but which do exhibit antimicrobial properties. Hydrogen peroxide is not commonly permitted but has been used as a temporary preservative especially of milk. Common salt is usually considered as an ingredient rather than as an additive. EDTA is normally addressed in legislation as a sequestering agent. BHA, BHT, TBHQ and propyl gallate are more commonly known as antioxidants and regulated as such.

Other substances which are antimicrobial include organic acids such as acetic acid, lactic acid, malic acid, etc., medium-chain fatty acids and esters, dimethyl dicarbonate and diethyl dicarbonate, antibiotics, gases such as carbon dioxide and sanitising agents such as chlorine. Various spices, herbs and vegetables contain antimicrobial agents which may be added to foods directly as the vegetable or may be extracted and added as a concentrate. Milk and eggs contain antimicrobials, e.g. lysozyme. Further information on these and other antimicrobials may be found in the reviews noted previously.

Tables follow on pp. 116–119.

References

Beuchat, L. R. and Golden, D. A. (1989) Antimicrobials occurring naturally in foods. *Food Technology* **43** (1), 134–142.

Branen, A. L. and Davidson, P. M. (eds) (1983) *Antimicrobials in Foods*. Marcel Dekker, New York.

Davidson, P. M. and Parish, M. E. (1989) Methods for testing the efficacy of food antimicrobials. *Food Technology* **43** (1), 148–155.

Daeschel, M. A. (1989) Antimicrobial substances from lactic acid bacteria for use as food preservatives. *Food Technology* **43** (1), 164–167.

ICMSF (1980) *Microbial Ecology of Foods*. Vol. 1: *Factors Affecting Life and Death of Microorganisms*. AP, New York.

Lewis, R. L. (1989) *Food Additives Handbook*. Van Nostrand Reinhold, New York.

Lueck, E. (1980) *Antimicrobial Food Preservatives*. Springer-Verlag, Berlin.

Roberts, T. A. (1989) Combinations of antimicrobials and processing methods. *Food Technology* **43** (1), 156–163.

Wagner, M. K. and Moberg, L. J. (1989) Present and future use of traditional antimicrobials. *Food Technology* **43** (1), 143–147, 155.

Williams S. (1984) *Official Methods of Analysis of the Accociation of Official Analytical Chemists*. (14th edn). AOAC, Arlington, VA, USA.

Table 5.1 Properties of commonly permitted antimicrobial preservatives

Preservative	Solubility at 25°C (g l^{-1})	Solubility of salts at 25°C (g l^{-1})	Yield of active form (%)	pH range (pK_a)[1]
Sorbic acid (2,4-hexadienoic acid) $CH_3CH=CH-CH=CH-CO_2H$	1.6 (water)	Sodium 320 (water)	83.6	(4.8)
	0.7 (10% saline)	Potassium 1380 (water)	74.6	
	130 (ethanol)	Potassium 540 (10% saline)		
	5–10 (edible oil)	Calcium 12.0 (water)	74.2	
Benzoic acid C_6H_5COOH	21.0 (water)	Sodium 660 (water)	84.7	(4.2)
		Calcium 40 (water)	75.8	
Parabens ($C_6H_4(OH)CO_2R$) where R = CH_3(methyl)	2.5 (water) 520.0 (ethanol)	Na readily water-soluble	87.4	(8.5)
C_2H_5(ethyl)	1.7 (water) 700.0 (ethanol)	Na readily water-soluble	88.3	
C_3H_7(propyl)	0.5 (water) 950.0 (ethanol)	Na readily water-soluble	89.1	
C_7H_{15}(heptyl), etc.	0.015 (water)	Na readily water-soluble	91.5	
Propionic acid $C_3H_6O_2$	Water-miscible liquid	Sodium 1000 Calcium 49	77.1 65.5	(4.9)
Nitrous acid HNO_2 (nitrites)	Readily water-soluble	Na readily water-soluble	68.1	(3.4)
		K readily water-soluble	55.2	
Sulphur dioxide SO_2 (sulphites)	110 (water)	K_2SO_3 250	33.0	$H_2SO_3 \rightleftharpoons$ $HSO^- + H^+$ (1.76)
		$Na_2S_2O_5$ 540	67.4	$HSO^- \rightleftharpoons$ $SO_3^{2-} + H^+$ (7.21)
		$K_2S_2O_5$ 250	57.6	
		Na_2SO_3 280	50.8	
		$KHSO_3$ 1000	53.5	
		$NaHSO_3$ 3000	61.6	
Nisin $C_{143}H_{230}N_{42}O_{37}S_7$	120 @ pH 2.5 (water) 40 @ pH 5.0 (water) water insoluble @ pH > 7.0	N/A	100	6.5–6.8

Table 5.1 *Continued*

Antimicrobial activity	Other effects	Uses in food (typical max. levels ppm)	(S)ynergists/ (A)ntagonists
Fungi – broad spectrum Bacteria – less effective (mostly strict aerobes) (not lactic acid bacteria)	N/A	Salad cream (1000) Salad dressing (1000) Mayonnaise (1000) Wine (200)	(A) Non-ionic surfactants (S) SO_2, CO_2, NaCl, sucrose, propionic acid, dehydration, nisin + polyphosphate
Yeasts and moulds Food poisoning bacteria Spore-forming bacteria Not spoilage bacteria	Phenolic taste	Soft drinks (160) Pickles (250) Fruit juice (800)	(A) Lipids – partitioning (S) SO_2, CO_2, NaCl, sucrose
Yeasts and moulds Bacteria – mainly Gram + Heptyl > propyl > ethyl > methyl	Phenolic taste	Similar foods in which benzoic acid is used, but active to higher pH (175)	(A) Proteins, emulsifiers, lipids, polysaccharides
Moulds but not yeasts *Bacillus mesentericus* (rope in bread)	Cheesy taste	Bread (3000) Flour confectionery (1000)	(S) CO_2, sorbic acid
At low Aw and + NaCl *C. botulinum* also (but less effective) *Lactobacillus, Bacillus, C. perfringens, Salmonella*	Colour development Flavour development Nitrosamines	Meat curing (200 ppm nitrite, 500 ppm nitrate)	(A) Sulphites, ascorbic acid (S) NaCl
Most bacteria (Less yeast and mould)	Reducing agent Bleaching agent Antioxidant Non-enzymic and enzymic browning inhibitor	Fruit juice (200) Sausages (450) Dried fruit and veg. (2000) Wine (450) Biscuit dough (70) Beer (70)	(A) Nitrous acid, many colours, many organics (sulphides, aldehydes, disulphides, thiols, isocyanates), vitamins, e.g. thiamine, lipids, sugars (S) benzoic acid, sorbic acid
Gram + bacteria, lactic acid bacteria, *Streptococcus, Bacillus, Clostridium*	N/A	Cheese (6–12 ppb) Dairy products (6–12 ppb) Canned foods (stable at low pH and high temp.) (6–12 ppb)	(A) High pH

Table 5.1 *Continued*

Preservative	Solubility at 25°C (g l^{-1})	Solubility of salts at 25°C (g l^{-1})	Yield of active form (%)	pH range (pK_a)[1]
Hexamethylenetetramine C$_6$H$_{12}$N$_{14}$ (hexamine)	Insoluble in water Soluble in most organics	N/A	Depends on degree of hydrolysis to formaldehyde	broad
Biphenyl C$_6$H$_5$C$_6$H$_5$	Insoluble in water Soluble in most organics	N/A	100	broad
2-hydroxybiphenyl (and sodium salt) C$_6$H$_5$C$_6$H$_4$OH	Insoluble in water Soluble in most organics	1220 (water) 280 (propylene glycol)	100 Na salt 88.5	broad
2(thiazol-4-yl)–benzimidazole C$_{10}$H$_7$N$_3$S	Insoluble in water Soluble in most organics	N/A	100	broad

[1] At pH above the pK_a, less than 50% of the active undissociated form is present.

Table 5.2 Properties of other antimicrobial preservatives

Preservative	Solubility at 25°C (g l^{-1})	Solubility of salts at 25°C (g l^{-1})	Yield of active form (%)	pH range (pK_a)[1]
Hydrogen peroxide H$_2$O$_2$	Readily soluble in water	N/A	100	Up to pH 10
Sodium chloride NaCl	N/A	311	100	Broad
EDTA disodium salt C$_{10}$H$_{14}$N$_2$O$_8$.2Na (ethylene diamine tetraacetic acid)	N/A	Freely soluble in water	100	Neutral to alkaline pH
BHA C$_{11}$H$_{16}$O$_2$ (butylated hydroxyanisole)	Insol (water) 400 (corn oil) 700 (propylene glycol)	N/A	100	4–9
BHT C$_{15}$H$_{24}$O (butylated hydroxytoluene)	Insol (water) 300 (corn oil)	N/A	100	4–9
TBHQ C$_{10}$H$_{14}$O$_2$ (tertiary butylhydroquinone)	<10 (water) 100 (corn oil) 300 (propylene)	N/A	100	4–9
PG C$_{10}$H$_{12}$O$_5$ (propyl gallate)	<10 (water) Insol (corn oil) 550 (propylene glycol)	N/A	100	4–9

[1] At pH above the pK_a, less than 50% of the active undissociated form is present.

Table 5.1 *Continued*

Antimicrobial activity	Other effects	Uses in food (typical max. levels ppm)	(S)ynergists/ (A)ntagonists
Hydrolyses to formaldehyde which is a broad spectrum antimicrobial	Formaldehyde binds to protein– causes toughening	Provolone cheese (25) Marinated fish (50) (Hydrolyses in acid and releases formaldehyde)	(A) High pH
Broad spectrum fungicide	N/A	Impregnated citrus wrap Citrus surface application (70) 1–5 g m^{-1} in impregnated wrappers	(S) Other fungicides
Broad spectrum fungicide	N/A	Citrus surface application (12)	(S) Other fungicides
Broad spectrum fungicide	N/A	Citrus surface application (3) Banana surface application (10)	(S) Other fungicides

Table 5.2 *Continued*

Antimicrobial activity	Other effects	Uses in food (typical max. levels ppm)	(S)ynergists/ (A)ntagonists
Most active vs. Gram − bacteria, e.g. coliforms. Intermediate vs. Staph. and lactic acid. Least vs. Gram +, e.g. sporeformers	Oxidant Bleaching agent	Preservation of milk (1000) (0.1 residual)	(A) Heavy metal ions (A) Catalase (A) Lactoperoxidase
Gram negative rods, e.g. *Pseudomonas* @ 4–10% Lactic acid bacteria @ 15%	Imparts salty taste	Butter (2%) Margarine (2%) Cheese (2%) Salted fish (28%)	(S) Refrigeration
Gram negative bacteria Clostridia	Chelating agent	Chelating agent (100–300) Surface application for fresh fish (100–300) Salad dressing (100–300)	(S) Sorbates vs. Staphylococcus (S) Nitrite vs. Clostridia (S) Lysozyme vs. Gram neg.
Most Gram + (*S. aureus, Bacillus, Clostridium*) Some Gram − (*Pseudomonas fluorescens, Vibrio*) Most fungi	Antioxidant	(200 of fat)	(S) Minerals, NaCl, other preservatives, acids (A) Unsaturated lipids, casein
C. botulinum *S. aureus*	Antioxidant	(200 of fat)	(S) Minerals, NaCl, other preservatives, acids (A) Unsaturated lipids, casein
B. subtilis *S. aureus* Most fungi	Antioxidant	(200 of fat)	(S) Minerals, NaCl, other preservatives, acids (A) Unsaturated lipids, casein
C. botulinum Most fungi	Antioxidant	(200 of fat)	(S) Minerals, NaCl, other preservatives, acids (A) unsaturated lipids, casein

6 Enzymes

Y. J. OWUSU-ANSAH

6.1 Introduction

Enzymes are inherently associated with biological materials such as food. They have, therefore, remained an important element in food processing. Various processes, dating back to antiquity have only been successful through the use of enzymes. To some extent one might not consider enzyme applications as science especially if one considers that some of the applications are being carried out successfully by non-scientists. For example in Ghana, West Africa, the local 'pito' is brewed by mostly illiterate brewers using enzymes inherent in sprouted sorghum. The pito brewer does not know that enzymes are responsible for the product and that he or she is applying science. To the trained scientist, the biochemistry of the process becomes apparent once the details are explained. The importance and the duration of use of enzymes in food applications probably explain why various books and book chapters (Reed and Underkofler, 1966; Whitaker, 1974; Ory and Angelo, 1977; Birch et al., 1981; Dupuy, 1982; Swaisgood, 1985; Bigelis and Lasure, 1987; Leiva and Gekas, 1988) and review articles (Ghildyal et al., 1979; Taylor and Richardson, 1979; Olsen, 1984; Sheppard, 1986; Ardeshir, 1987; Alder-Nissen, 1987; Olsen and Christensen, 1987; Klacik, 1988) have been written on enzymes in food processing. Inherent biochemical processes such as fermentation, milk clotting and tenderization of meat are due to the action of these biocatalysts previously, and erroneously thought to be solely proteinacious. In some of these applications the enzymes are associated with the cell as compared to addition of exogenous commercial enzymes. Present enzyme technology allows better quality enzymes than those available by traditional processes to be produced. The use of these enzymes improves processes (Madsen et al., 1973; Kilara, 1982; Law and Wigmore, 1982) reduces cost (Jacobsen and Rasmussen, 1983) and even allows new and innovative products to be made (Aldler-Nissen, 1977; Gunther, 1979; Strobel et al., 1983). Although there are various food applications where enzymes associated with the cells are still used, these applications are not discussed in this chapter. Interested readers are referred to other excellent sources such as *Source Book of Food Enzymology* (Schwimmer, 1981) and *Enzymes in Food Processing* (Reed and Underkofler, 1966). The purpose of this chapter is to provide to the practising food scientist or technologist with practical information on application of commercial enzymes in food processing and technology.

6.2 Basic enzymology

6.2.1 Enzyme nomenclature

The nomenclature of enzymes has evolved over several years. In the scientific world systematic names developed by the International Union of Biochemistry Committee on Nomenclature and Classification of Enzymes is used. The committee grouped enzymes into six major classes based on the type of reaction they catalyse. The main six groupings are: (1) oxidoreductases; (2) transferases; (3) hydrolases; (4) lyases; (5) isomerases; (6) ligases. In this system of nomenclature an enzyme is identified by the prefix EC followed by a series of numbers with periods after each number. The first number refers to the main group designation as described above. The other series of numbers describe the specific subclasses and subsubclasses of the enzyme according to the nature of the reaction they catalyse. Another form of nomenclature, recommended by the International Council, are those referred to as trivial names. By this nomenclature enzymes are named by adding the suffix 'ase' to the name of their substrate, e.g. lipid hydrolysing enzymes are known as lipases. An additional nomenclature such as adding an 'in' suffix to the name of the source of the enzyme is also practised. A common example is the enzyme papain which is extracted from the papaya plant. The systematic names are quite complex, very long to use or write, therefore, practical technologists or scientists conveniently use the trivial names. For brevity the trivial names of the enzymes are used in this chapter. Also the enzymes have conveniently been classified into general groupings; depending on the general reaction they catalyse (e.g. carbohydrase, protease, etc.) Where there could be confusion, the trivial as well as the systematic nomenclature have been provided.

6.2.2 Factors modulating enzymic reactions

Various factors affect enzyme activity and hence the rate of the enzyme-catalysed reactions. Some of these factors, which if properly understood could also be exploited to the advantage of the scientist or technologist, are: (1) concentration and availability of substrate, (2) concentration of enzyme, (3) temperature, (4) pH of reaction, (5) residence time of reaction and (6) presence or absence of inhibitors or activators.

Generally at low substrate concentrations enzymic reactions follow a zero-order kinetics. In food applications, the enzymes are present in minute levels thus the substrate concentration is very high compared to that of the enzyme. The reaction rate of most such enzymic processes follow first-order reaction kinetics:

$$\partial P/\partial T = k(S-P)$$

where $\partial P/\partial T$ is the rate of product formation and $(S-P)$ is the concentration of substrate remaining at a given time. The reaction rate is therefore dependent on the amount of substrate remaining. An important practical use of substrate concentration is to use it to improve the stability of the enzyme. Enzymes denature more rapidly in dilute solutions. At high substrate concentrations the catalytic sites of the enzymes are saturated enhancing stability. Substrate availability is very important in practical enzymology. The mere fact that a substrate is available in high concentrations does not necessarily mean that the reaction would proceed. Enzymes and substrates that are compartmentalised will not react. The substrate must be in contact with the enzyme for the catalytic function to proceed. A typical example in food application is the hydrolysis of triacylglycerol with lipases. The enzyme is hydrophilic while the substrate is lipophilic. The use of an emulsifier considerably increases the hydrolytic rate by bringing the enzyme and substrate together.

The enzyme concentration is always relatively small. For most food applications the reaction rates are proportional to the enzyme concentrations. Exceptions to this occur if reactions are left to proceed to very low substrate levels.

Temperature affects enzymes in two ways; acceleration or retardation of the catalysed reaction with increasing temperature, culminating in the complete denaturation of the enzyme as the temperature exceeds the critical denaturation point of the enzyme. Most chemical reactions proceed faster as the temperature is raised apparently due to increase in kinetic energy of the reactants. Enzymes behave similarly until a critical temperature is reached where the rate starts to decrease. The temperature at which an enzymic reaction is carried out is therefore of practical importance. At higher temperatures the rate of reaction and enzyme denaturation would equally be high. In most cases a compromised optimum temperature is selected for use. Temperature is also used to stabilise enzymes against other factors. For example by lowering the reaction temperature the catalytic pH range of an enzyme can be expanded and, therefore, stabilised for a particular reaction.

The effect of pH on enzyme activity is profound. Some enzymes show a very broad optimal pH range while others are very sharp. The stability of an enzyme is also affected by pH. Even localised pH changes can adversely affect an enzymic reaction. In production situations it seems always prudent to start adding the enzymes to the reaction vessel when the pH is at or close to the required level. Also acids or alkaline metering points should be remote from where the enzyme is being added to the reaction mixture.

Time is obviously of the essence in enzymic reactions since time is needed for the enzyme to react with the substrate. In first-order enzyme-catalysed reactions, the reaction velocity decreases with time, i.e. as the available substrate diminishes. Such enzyme-catalysed reactions require considerable amount of time to attain completion.

Chemical compounds termed inhibitors can adversely affect enzymic reactions. Such compounds could be metals such as copper, iron or calcium or they could be from the substrate itself. An example of substrate associated inhibitors is trypsin inhibitor from soybean. Depending on the mode of action of the inhibitor they could be 'competitive', 'non-competitive', 'uncompetitive', 'mixed-type', 'specific' or 'non-specific'. Some compounds may act as enzyme activators either by improving the stability of the enzyme or by enhancing its catalytic activity. The details and kinetics of enzymic inhibitors and activators are outside the scope of this chapter. Readers interested in that information are referred to biochemistry or enzymology textbooks.

6.2.3 Enzyme assay

Methods for determining the activity of enzymes are very important in both enzymology and enzyme applications and have been the subject of various publications (Colwick and Kaplan, 1955; Underkofler et al., 1960; Barman, 1969; Bergmeyer, 1974). The molar concentrations of most enzyme preparations are unknown, therefore, the amount of enzyme present is expressed in terms of its activity. In most cases enzyme manufacturers present information on the activity of their preparations based on standard analytical conditions. Enzyme activity is frequently expressed in the International Unit (IU) which is defined as the amount of the enzyme that catalyses the formation of 1 mmole of product per minute under defined conditions. These 'defined conditions' allow assays to be performed at different pH, temperature, etc., since these factors affect enzyme activity.

Another unit of enzyme activity recommended by the International Union of Biochemistry Commission on Enzymes is the katal. One katal is defined as that amount of enzyme that catalyses the conversion of one mole of substrate per second. One International Unit is thus equivalent to 16.67 katal. Most enzymic activities are rarely expressed in katals despite this unit being in conformity with the SI Unit.

Enzymes are considered as food additives in some countries and are, therefore, regulated. Most regulatory agencies such as the Food and Drug Administration of the United States of America (FDA) use specific methods to test activity of enzyme preparations. The FDA for example uses the procedures recommended in the Food Chemical Codex published by the Food Nutrition Board of the National Research Council of the National Academy of Sciences in Washington DC (NAS/NRC, 1981). The assays are specific for various enzymes approved for use in foods. The procedures are used by both national and international food regulatory agencies. Readers are recommended to consult this reference for details.

The procedures in Food Chemical Codex cover considerable numbers of enzymes used in food processing although they do not include assays for new uses of some of these enzymes. For example the assay for lipase in the Food Chemical Codex is based on the hydrolysis of olive oil under some standard conditions. This is acceptable if the lipase is being used for triacylglycerol hydrolysis. If, on the other hand, the enzyme is being used to catalyse the synthesis of triacylglycerol or interesterification of ester groups in an oil, the method in the Food Chemical Codex cannot be used to follow such reactions. For this reason a summary of generalised methods for use in most practical applications have been summarised in Table 6.1. Interested readers may consult the references provided in the summary for details of the procedures.

6.3 Choosing an enzyme for food application

Several factors need to be taken into consideration when choosing an enzyme for food application. Obvious among these factors are:

1. *The source, form and legal status of the enzyme.* It is important that the enzyme of choice and its source are permitted for use in the specific application.
2. *Availability of supply of consistent quality.* It is worth while ensuring that an acceptable quality and consistent supply is guaranteed by the supplier.
3. *Convenience of usage; immobilised or soluble enzymes.* Some enzyme preparations are in solutions, others in powders and/or immobilised systems. Most food applications of enzymes employ soluble liquids or powders, basically due to the nature of the applications. However, most of the widely marketed enzyme preparations are in immobilised forms.

    Enzyme immobilisation was first reported about 74 years ago (Nelson and Griffin, 1916); however, most developments on the technology have occurred within the last two decades. The techniques for enzyme immobilisation vary considerably but the general approaches used are:

    (a) Covalent bonding of the enzyme onto a solid or insoluble support material.
    (b) Adsorption of the enzyme on a support through non-specific and/or

Table 6.1 Summary of practical assay methods for enzymes used in foods

Enzyme class	Assay method and principle	Major applications	References
Carbohydrases	1. *Iodine test.* Iodine reacts with polymers containing over six glucose molecules in 1–4 bonds to give a blue colour-complex. Non-reactive to cellulose.	Determine the completion of starch hydrolysis	ASBC Method 8
	2. *Copper reduction.* Aldehydic sugars reduce copper ions. The reduced copper can then be measured quantitatively by various means; gravimetric, spectrophotometric or titration.	Determine the extent of hydrolysis of carbohydrates to sugars	
	3. *Polarimetry.* The plane of polarized light incident on sugar solutions is rotated. The extent and direction of rotation is related to the concentration and type of sugar and can, therefore, be used to follow enzyme reactions where one type of sugar is produced.	To determine extent of hydrolysis of starch to glucose	Bates (1942)
	4. *Osmometry.* The colligative properties of a solution such as freezing point, is affected by the concentration of the solution. The change in these properties can be used to determine the concentration of the solutes.	Follow hydrolysis of biopolymers such as starch, pectic substances, etc.	Fitton (1979); Nijpel *et al.* (1980)
	5. *HPLC.* The relative affinity of column packing materials to compounds can be used to chromatographically follow enzymic reactions. Normal phase and ion-exchange separations are possible.	Application varies considerably	Engle and Olinger (1979)
	6. *Viscosity.* A solution of a polymer exhibits some resistance to flow when a force is applied to it. The reduction in this resistance to flow due to enzymic breakdown can be used as an assay for the enzyme activity.	For following enzymic breakdown of biopolymers	AACC #22-10
	7. *Enzymic.* Enzymes such as glucose oxidase react with a substrate to stoichiometrically release products that can be quantified. The measure of the products gives indications on the concentration of the substate.	Used to determine presence of glucose in presence of other sugars	Food Chemical Codex (NAS/ NRC, 1981); Scott (1953)

Table 6.1 *Continued*

Enzyme class	Assay method and principle	Major applications	References
	8. *Turbidity.* Some carbohydrate polymers can be precipitated by addition of alcohol. The turbidity due to alcohol precipitation is related to the concentration of the polymer in solution.	Used to follow hydrolytic breakdown of polymers that can be precipitated with alcohol, e.g. pectin	AACC #22-07
Lipases (FFA)	1. *Titration.* Production of free fatty acids (FFA) during hydrolysis of fat yield H^+ that can be titrated with a common base. The concentration of H^+ produced indicates the activity of the enzyme.	Most common assay for lipase activity in laboratories based on FFA production	Jensen (1983); Maskowitz *et al.* (1977)
	2. *pH–Stat Method.* Principle is based on FFA production which tends to lower pH. Amount of base added to restore pH is a measure for the enzyme activity.	FFA for enzymatic reactions	Food Chemical Codex (NAS/ NRC, 1981)
	3. *Acid-value.* Based on titration of sample in a mixture of isopropanol and toluene using KOH to pH 7.0.	For FFA in fats and oils	AOCS Method Cd 3a-63
	4. *Dole extraction.* FFA is extracted under acid conditions from a sample into a heptane/ isopropanol mixture. The extracted solution is titrated to determine the FFA present.	For precise determination of FFA in fats and oils	Dole and Mienertz (1960)
	5. *Folch extraction.* FFA in sample is repeatedly extracted in acidified mixture of dichloromethane and methanol. After evaporating the solvent, the extracted oil is titrated to determine the amount of FFA.	For precise determination of FFA in fats and oils	Folch *et al.* (1957)
	6. *Silicic acid titration.* An acidified sample is mixed with silicic acid and the powder extracted with a mixture of dichloromethane and methanol. The filtrate is titrated to determine the FFA.	Used for FFA determination in liquid materials containing non-lipid materials	Sampugna *et al.* (1967)
Lipases (FFA) (acylglycerols)	1. *AOCS–α-monoglyceride content* (α-MG). The α-MG is extracted and oxidised with periodic acid. The α-MG is then back titrated.	Used for following α-MG formed during glycerolysis.	AOCS Method No. Cd 11-57; Sonntag (1982)

Table 6.1 *Continued*

Enzyme class	Assay method and principle	Major applications	References
	2. *Triacylglycerol (TG) isolation.* Dissolved sample in heptane is loaded onto an activated alumina column which is then selectively eluted with diethyl ether. After evaporation the TG can be estimated gravimetrically.	Total TG in most fat systems, except emulsified systems when prior extraction of the fat is recommended	Novo industries Method No. F-840289
Lipases (lipid profile)	1. *AOCS method.* FFA are esterified to make them volatile. The converted FFAs are analysed on a GLC.	For FFA profile determination of animal and vegetable oils	AOCS Method No. Ce-1-62 (1981)
	2. *Dairy products profile.* The FFA in the sample are extracted with 1% butanol in petroleum ether after acidification with a mixture of sulfuric acid and silicic acid. The profile is determined by GC.	For profiling FFA of dairy products	Woo and Lindsay (1982)
	3(a) *HPLC method.* Preferential affinity of different FFA for column packings allows separation of various FFA on an HPLC reverse phase column.	For profiling interesterified oils	Jensen (1981); Shulka *et al.* (1983)
	(b) Lipid class samples can also be resolved using an RPC–HPLC.	Used to profile lipid classes for animal and vegetable oils	Payne-Wahl *et al.* (1981)
	4. *TLC method.* Extracted from samples or prepared in organic solvents. Aliquots are spotted on silica gel TLC plates. The lipid classes are separated in a solvent system and the plate developed with cupric acetate and charred. The resolved spots are quantified using a densitometer.	For resolving lipid classes in various commodities of animal and vegetable origin	Bitman and Wood (1981)
Proteases	1. *Free amino nitrogen method.* Hydrolysis of proteins yields free amino group that can react with various chromogenic agents. The intensity of the chromogen complexes provides a measure of the free amino group and hence extent of hydrolysis.	For determining precisely the extent of hydrolysis of a substrate by a protease	Adler-Nissen (1979); Moore and Stein (1948)
	2. *Gel permeation chromatography.* This is a mode of chromatography which separates molecules based on their size or molecular weight (MW).	To follow extent of hydrolysis and characterise hydrolytic products based on their MW	Regnier (1983)

Table 6.1 *Continued*

Enzyme class	Assay method and principle	Major applications	References
	3. *Reverse phase peptide mapping.* Peptides, proteins are separated based on their polarity and affinity to chromatographic column packing.	For identifying well-characterized hydrolytic and substrate materials	Fullmer and Wasserman (1979)
	4. *Viscosity.* Reduction in viscosity of a solution of protein on enzyme treatment is used to assay for protease activity.	Suitable for use whenever a change in viscosity is associated with the enzymic reaction	Dealy (1984); Kupier et al. (1978)
	5. *pH Stat/pH Drop.* Hydrolysis exposes amino acid side groups. At alkaline or neutral pH the pH of the medium drops. The amount of base to restore the pH is related to the extent of hydrolysis as follows: $$DH = \frac{B_v \times N_b}{M_p} \times \frac{1}{\alpha} \times \frac{1}{H} \times 100\%$$ where DH = degree of hydrolysis B_v = volume of base to restore the pH N_b = normality of base used M_p = mass of protein in reaction $1/\alpha$ and $1/H$ are constants	For practically following extent of hydrolysis for protease reactions conducted at pHs higher than 6.5	Alder-Nissen (1977)
	6. *Osmometry.* The depression of freezing point or vapour pressure with increase in concentration of solute is applied here. As proteolysis proceeds, the amount of soluble protein increases. This increase tends to alter the colligative properties of the solution. The depression in freezing or vapour point created by the increase in concentration is used to monitor the enzyme activity.	A quick way to follow protein hydrolysis in plant or pilot plant situations	

Compiled from Boyce (1986).

specific bonding forces, such as electrostatic, hydrophobic interactions or affinity bonding.

(c) Cross-linking of the enzyme to itself by reacting with a bifunctional reagent.

(d) Entrapment of the enzyme within a matrix in which the enzyme would be accessible to the substrate.

(e) Encapsulation of the enzyme in a semipermeable membrane capsule.

The use of immobilized enzymes offers tremendous advantages over soluble ones. Most commercially successful immobilised enzymes show equal or better stability than soluble ones. Immobilisation allows easier process control, removal of enzyme from product, reusability and hence reduction in cost. It also allows continuous processes to be developed. There are also a few drawbacks on immobilised enzyme systems: substrate accessibility could be diminished significantly, immobilised support systems could be breeding grounds for microorganisms and there may be problems of waste disposal of spent immobilised preparations.

Despite the advantages that immobilised enzymes seem to offer there are not many commercially available immobilised enzymes possibly because most of the enzymes cannot be effectively immobilised using the presently known technologies. Table 6.2 shows a list of commercially available immobilised enzymes used in food processing. When faced with choosing between immobilised and soluble enzymes for an application in which both are suitable, it seems to always make economic sense to choose the immobilised system.

Table 6.2 Commercially available immobilised enzymes used in food processing

Enzyme	Immobilisation methods	Product	References
Amino acylase	Adsorption	L-Amino acid	Ghildyal et al. (1979); Poulsen (1984); Swaisgood (1985)
Amyloglucosidase	Adsorption	Glucose	Poulsen (1984)
	Entrapment	Glucose	Swaisgood (1985)
	Cross-linked	Glucose syrup	Swaisgood (1985)
Glucose Isomerase	Cross-linked	High fructose syrup	Poulsen (1984)
	Adsorption	High fructose syrup	Swaisgood (1985)
	Entrapment	High fructose syrup	Swaisgood (1985)
α-Galactosidase	Entrapment	Sucrose	Swaisgood (1985)
β-Galactosidase	Covalent cross-linking	Glucose and galactose	Poulsen (1984)
	Adsorption	Glucose and galactose	Poulsen (1984); Swaisgood (1985)
	Entrapment	Glucose and galactose	Poulsen (1984); Swaisgood (1985)
Lipase-Esterase	Adsorption	Terpene esters Triacylglycerols Free fatty acid	
Tannase	Adsorption	Tea tannins in instant products	Godfrey (1983)

4. *Cost*. The cost of the enzyme is very important in choosing an enzyme; especially in deciding on enzyme preparations from different suppliers. It is always advisable to compare the quotations from suppliers based on cost per unit activity of the enzyme. Most suppliers provide technical data sheets for their products which normally have the specific activities of the preparations. Comparing two such specific activities of two preparations could be erroneous since the procedure of assay used by the suppliers could be different. The scientist or technologist should ensure that the enzyme activities being compared are based on the same assay procedures. Where there are uncertainties it is always advisable to determine the activities in the laboratory and compare the cost based on the laboratory results.

6.4 Food applications

The uses of enzymes in the food industry cover a wide range of applications including synthetic and degradative functions. The choice of an enzyme for a specific application often takes into consideration the source and the biochemical characteristics of the enzyme. The source is an important factor since for food applications it is regulated in some countries. Table 6.3 summarises the sources and biochemical characteristics of the enzymes used in food processing. In Table 6.4 the specific applications of the enzymes and their levels of use are presented.

6.5 Legislation and regulation of enzymes used in foods

Similar to other additives used in foods, the use of enzymes in food is regulated by legislation. The legislative requirements and interpretations, however, vary from one country to another. The regulations existing in most countries are dynamic; undergoing constant revisions when necessary. It is important to know the regulations of a specific country for which an enzyme-processed product is going to be marketed. In some cases the regulations in the country would not only determine the way the food is manufactured, but may even affect the labelling on the packaging of the food.

Most developing countries follow the Joint FAO/WHO Expert Committee on Food Additives Regulations. There is no common agreement in the European Community countries. In Belgium and Italy enzymes are considered as processing aids and their use is subject to authorisation. In Greece, Ireland, The Netherlands and the United Kingdom (UK), there are no specific controls on enzyme use. In the UK, however, the source of the enzyme should be on the permitted list of the regulation agencies (Table 6.5).

Table 6.3 Sources and biochemical characteristics of enzymes used in food processing

Enzyme type	Classification	Source	Optimum values		Observations	Side activities
			pH	Temp. (°C)		
α-Amylase	1,4-α-D-glucan glucanohydrolase 3.2.1.1	Cereals	5–6	50	Activated by calcium ions Inhibited by oxidising agents	β-Amylase, β-glucanase, neutral acid protease
	Endo-hydrolysis of polysaccharides with 3 or more 1,4-α-D-glucan units	Pancreas	6.5	40	Calcium salts increase heat stability	Esterase, lipase and protease
		Aspergillus niger	5	55	Activated by calcium ions	Cellulase, hemicellulase, acid protease, xylansase
β-Amylase	1,4-α-D-glucan maltohydrolase 3.2.1.2	Cereals	5.5	55	Activated by reducing agents	α-Amylase
	Exo-removal of maltose units from non-reducing end of polysaccharide chains	Soya bean	4–7	55	Acid tolerant	α-Amylase, lipoxygenase
		Bacillus sp.	5–7	60	—	α-Amylase, β-glucanase, neutral protease
Amyloglucosidase	1,4-α-D-glucanohydrolase 3.2.1.3	Aspergillus oryzae	4–5	55	—	Also hydrolyses α-1,6 bonds in the starch
		Aspergillus niger	4–5	55		
		Rhizopus sp.	4–5	55		
		Trichderma viridue	—	—		
		Aspergillus oryzae	5	55	Activated by calcium ions	Glucoamylase, acid protease
		Bacillus subtilis	6–7	70–80	Activated by calcium ions Inhibited by chelating agents	β-Glucanase, acid and neutral protease

Table 6.3 *Continued*

Enzyme type	Classification	Source	Optimum values		Observations	Side activities
			pH	Temp. (°C)		
		Bacillus licheniformis	7–8	90–95	Low calcium dependence, especially in presence of high substrate	β-Glucanase, acid and neutral protease; these are thermolabile and rapidly inactivated at amylase use temperatures
Anthocyanase (Okada et al., 1968)	Anthocyanin-β-glucosidase	*Aspergillus niger*	3–9	50		β-Glucosidase activity
Catalase	Hydrogen peroxide; hydrogen peroxide oxido-reductase 1.11.1.6	*Aspergillus niger* / Bovine liver	5–8 / 7	35 / 45	Stable at low pH / Inactivated by alkali	Usually very pure / Usually very pure
Cellobiase	β-D-glucoside glucohydrolase 3.2.1.21 Exo-hydrolysis of terminal non-reducing 1,4-α-D-glucose residues	*Aspergillus niger*	5	60	Used to reduce product inhibition of cellobiose when cellulases used	Amylase, glucoamylase, protease, hemicellulase
Cellulase	1,4-(1,3; 1,4)-β-D-glucan 4-glucanohydrolase 3.2.1.4	*Aspergillus niger*	5	45	Generally low in C_1 type activity	Amylase, cellobiase, glucosidase, glucoamylase-

	Source	pH	Temp.	Remarks	Associated enzymes
Endo-hydrolysis of 1,4-β-glucosidic links of cereal glucans, cellulose, lichenin	Basidiomycetes sp.	4	50	Good C_1 type with broad specificity	Hemicellulase, protease
	Penicillium funiculosum	5	65	Product inhibition is usually low	Amylase, glucoamylase, cellobiase
	Rhizopus sp.	4	45	Broad specificity	Amylase, glucoamylase, protease
	Trichoderma sp.	5	55	High C_1 activity	Hemicellulase
Dextranase 1,6-α-D-glucan 6-glucanohydrolase 3.2.1.11 Endo-hydrolysis of dextrans	Penicillium sp.	5	55	Products are isomaltose and isomaltotriose	Cellulase, hemicellulase
Diacetyl reductase Anthocyanin-β-glycosidase	Aerobacter aerogenes	6–8	30	Inhibited by ethanol NADH activated.	—
α-Galactosidase α-D-galactoside galactohydrolase 3.2.1.22	Aspergillus niger	4.5	65	Acts on many galactosides	Glucosidase, hemicellulase
Exo-hydroglysis of terminal non-reducing α-D-galactoside residues of polysaccharides, oligosaccharides, galactomannans, galactolipids	Saccharomyces sp.	5	50	Low galactomannanase activity	Glucosidase, invertase
β-Galactosidase β-D-galactoside galactohydrolase 3.2.1.23	Aspergillus niger	4.5	55	Product inhibition and the presence of transferase activity is common to all 5 types	α-L-arabinase, glucanase, glucosidase, transferase, invertase, acid protease all common to both Aspergillus sp.
Hydrolysis of terminal non-reducing β-D-galactose residues	Aspergillus oryzae	4.5	55		
	Bacillus sp.	7.3	60		Amylase, glucosidase, protease,

Table 6.3 *Continued*

Enzyme type	Classification	Source	Optimum values		Observations	Side activities
			pH	Temp. (°C)		
		Kluyveromyces sp.	6.5	45		Glucosidase, invertase, protease, transferase common to the yeasts
		Saccharomyces sp.	6.5	40		
β-Glucanase	Endo-1,3(4)-β-D-glucanase 3.2.1.6	*Aspergillus niger*	5	60	Broad range with low specificity	Amylase, glucoamylase, glucosidase
	Endo-hydrolysis of β-D-glucans when the glucose residue whose reducing group is involved is itself substituted on carbon 3	*Bacillus subtilis*	7	50–60	Narrow range with higher specificity	Amylase, glucosidase, protease
		Penicillium emersonii	4	70	Tolerant of low pH in some variants	Amylase, dextranase, protease
Glucoamylase	1,4-α-D-glucan glucanohydrolase 3.2.1.3 Exo-hydrolysis of terminal 1,4-α-D-glucose residues from noreducing end of polyglucoside chains	*Aspergillus awamori*	4–5	60	Acid tolerance and thermotolerance vary widely from source to source	Amylase, glucanase, cellulase, hemicellulase, protease; all common to these fungal sources
		Aspergillus niger	3–5	65		
		Aspergillus oryzae	4.5	60		
		Rhizopus sp.	2.5–5	55		
Glucose isomerase	D-xylose ketol isomerase 5.3.1.5	*Atinoplanes missouriensis*	7.5	60	In the immobilised form, all these are activated by magnesium and cobalt; need for	All are usually pure
	A true xylose isomerase acting	*Bacillus coagulans*	8	60		

Enzyme	Systematic name and EC number	Source	pH	Temp (°C)	Characteristics	Other enzymes present
	on glucose at high substrate concentration	*Streptomyces* sp.	8	63	cobalt varies with preparation; magnesium competed for by calcium and must be in excess	
		Streptomyces albus	6–7	60–75		
Glucose oxidase	β-D-glucose; oxygen 1-oxidoreductase 1.1.3.4	*Aspergillus niger*	4.5	50	—	Catalase
		Aspergillus sp.	2.5–8	15–70	Acid tolerance, thermotolerance among gp.	Catalase
		Penicillium notatum	3–7	50	—	Catalase
β-Glucosidase	β-D-glucoside hydrolase 3.2.1.21	*Aspergillus niger*	5	60	Used to reduce cellobiose inhibition on cellulase	Amylase, glucoamylase, protease
		Aspergillus oryzae	5	65	Broad specificity, unusually high thermotolerance	Amylase, glucoamylase, protease
		Bacillus sp.	7	70	Narrow specificity	Amylase, glucanase, protease,
		Clostridium thermocellum	9	60	Very active on β-1,3-bonds	Protease
		Saccharomyces sp.	7	45	Broad specificity	Glucanase, protease
		Sweet almond	7	50	Broad specificity	Usually very pure
		Trichoderma viride	5	65	Broad specificity	Hemicellulase
Hemicellulase	Many activities repesented; examples include: Endo-1,4-β-D-Mannan hydrolase 3.2.1.78 Exo-α-L-Arabino-furan hydrolase 3.2.1.55	*Aspergillus* sp.	3–6	70	These are very complex enzyme systems that contain many activities that require specific substrates to identify and distinguish them	In common with the *Aspergillus* group, cellulase, glucosidase, pectinase, pentosanase

Table 6.3 *Continued*

Enzyme type	Classification	Source	Optimum values		Observations	Side activities
			pH	Temp. (°C)		
Inulinase	Exo-1,3-β-D-Xylan hydrolase 3.2.1.72					
	2,1-β-D-Fructan fructanhydrolase 3.2.1.7	*Aspergillus* sp.	4.5	60	—	Amylase, glucoamylase, invertase, protease, can be anticipated in either type
	Endo-hydrolysis of inulin	*Candida* sp.	5	40	—	
Invertase	β-D-Fructofuranoside fructohydrolase 3.2.1.26	*Candida* sp.	4.5	50	Maximum activity is shown at low substrate levels	In common with other yeast derived enzymes, the proteases may be present
		Saccharomyces sp.	4.5	55	All contain a bound mannan	
Lactase	*See* β-Galactosidase					
Lipase	Triacylglycerol acylhydrolase 3.1.1.3	*Aspergillus niger*	5–7	40	High in esterase	Amylase, cellulase, esterase, hemicellulase, pectinase, protease
		Candida cylindracea	8	50	Active on higher oils and fats	Esterase, protease
		Mucor miehei	7.5	50	High in true lipase activity	Esterase, protease
	Fatty acid esterase Carboxylic-ester hydrolase 3.1.1.1	Pancreatic	7.5–8	40	Preferential action on triacylglycerols	Amylase, protease

		Source	pH	Temp. (°C)	Notes	Side activities
	Aryl-ester hydrolase 3.1.1.2	Pregastric esterase	5.5–7	30–60	High esterase–lipase ratios	Amylase
		Rhizopus sp.	5–8	40	Very varied specificities	Amylase, cellulase, esterase, protease
Metallo–neutral	Microbial 3.4.24.4 Preferential cleavage; bonds hydrophobic residues	Aspergillus oryzae	7	50	All are inhibited by reducing agents, chelating agents, halogens	As for most protease, the side activities are other proteases
		Bacillus thermoproteolyticus	8	65		
		Bacillus sp.	7	50		
Naringinase	β-L-1-rhamnosidase 3.2.1.40	Penicillium sp.	3–5	40	—	β-glucosidase activity (3.1.21)
Pectinase	Poly(1,4-α-D-galacturonide) glycanohydrolase 3.2.1.15	Aspergillus sp.	2.5–6	40–60	Wide variety of component activities	Pectin, esterase, pectin lyase, etc.
		Rhizopus sp.	2.5–5	30–50		
Penicillin amidase	Penicillin amidohydrolase 3.5.1.11	Bacillus sp.	7–8	37	—	Usually very pure
		Basidiomycetes sp.	4–6	50	—	Usually very pure
Peroxidase	Donor; hydrogen peroxide oxidoreductase 1.11.1.7	Horse radish	5–7	45	A specific haem protein enzyme	Catalase
Proteases Serine types	Chymotrypsin 3.4.21.1 Preferential cleavage of Tyr, tryptophan, Phe, Leu residues	Pancreatic	8–9	35	Inhibited by compounds in cereals, beans, potato, egg. Bovine more thermotolerant	Amylase, lipase, esterase

Table 6.3 *Continued*

Enzyme type	Classification	Source	Optimum values		Observations	Side activities
			pH	Temp. (°C)		
	Trypsin 3.4.21.4 Preferential cleavage of Arg and Lys	Pancreatic	8–9	45	Inhibitors as for Chymotrypsin Bovine more thermotolerant	Amylase, lipase, esterase
	Subtilisin 3.4.21.14 Hydrolysis of proteins peptide	*Bacillus amyloliquefaciens*	9–11	60–70	All are subject to inhibition by organo-phosphorous compounds	All are likely to show amylase, glucanase and additional protease
		Bacillus licheniformis	9–11	60–70		
		Bacillus subtilis	9–10	55		
		Aspergillus oryzae	8–10	60–70	No peptidase activity	Amylase, glucoamylase cellulase, hemicellulase
Thiol types	Papain 3.4.22.2 Preferential cleavage of Arg; Lys	*Papaya sp.*	5–7	65	All are activated by reducing compounds All are inhibited by oxidizing agents	Lysozyme, glucanase, glucosidase, cellulase
	Ficin 3.4.22.3 Preferential cleavage of Lys; Ala; Tyr; Gly; Asn; Leu; Val	*Ficus sp.*	5–7	65		Lysozyme, esterase, peroxidase
	Bromelain 3.4.22.4 Preferential cleavage of Lys; Ala; Tyr; Gly	*Ananas sp.*	5–8	55		Cellulase, hemicellulase, pectinase

Carboxyl-acid types					
Pepsin 3.4.23.1 Preferential cleavage of Phe; Leu	Porcine mucosa	1.8–2	40–60	Inhibited by aliphatic alcohols	Usually very pure
Chymosin (Rennet) 3.4.23.4 Specific for 1 bond of kappa casein	Bovine abomasum	4.8–6	30–40	—	Usually pure but may contain pepsin
Pullulanase					
Pullulan-6-glucano hydrolase 3.2.1.41 Hydrolysis of 1,6-α-D-glucosidic link in pullulan, amylopectin, glycogen and limit dextrins	Bacillus sp.	4.5	60	—	Protease, amylase
	Klebsiella aerogenes	5	50	—	Protease, amylase
Tannase					
Tannin acylhydrolase 3.1.1.20	Aspergillus niger	4.5	55	—	Amylase, glucoamylase
	Aspergillus oryzae	3–5	45	—	Amylase, glucoamylase, protease, cellulase
Xylanase					
1,3-β-D-Xylan xylanohydrolase 3.2.1.32	Aspergillus niger	3–5	45–55	—	Amylase, glucoamylase, glucosidase, cellulase
Endo-hydrolysis of 1,3-β-D-xylose units of xylans	Aspergillus oryzae	4	45	—	Amylase, glucanase, glucoamylase, cellulase
	Bacillus sp.	7–9	55–65	—	Amylase, glucanase protease

Adapted from Godfrey and Reichelt (1983).

Table 6.4 Enzyme applications in food processing. GOU – Glucose oxidase unit; DS – Dry solids; E/S – Enzyme/substrate ratio

Food group	Enzyme	Specific application	Mode of use	Dosage level or operational life
Cereals and starches	Amylases (Sheppard, 1986)	*Baking* Acceleration of fermentation. Improve bread flour to yield loaves of increased volume, improve crust colour and crumb structure	Free liquid form or tablets	0.002–0.006% of the flour
	Proteases (Reed and Underkofler, 1966)	Modification of gluten in baking of biscuits and to reduce mixing time of dough	Powders	up to 0.25% of flour
			Tablets	75HU per 100 g flour
	Pentosanase	Improvement of dough handling and bread quality	Powders	—
	Amylases (Sheppard, 1986)	*Starch liquefaction* Reduction of maltose	Liquid, for jet cooking	0.05–0.07% DS
			Enzyme/Enzyme liquid	0.05–0.1% DS
	Amyloglucosidase (Sheppard, 1986)	Production of glucose	Liquid with syrup or without other enzymes	0.06–0.131% DS
	Glucose Isomerase	*Conversion of Glucose to Fructose*	Immobilised	0.015–0.03% DS Fixed bed 0.16 DS Batch
Alcoholic beverages	Amylases (Sheppard, 1986)	*Brewing* To reduce viscosity of mash	Liquid	0.025%
		Conversion of starch to sugars for fermentation	Liquid	0.003%
	Tannase	Removal of polyphenolics	Liquid or powder	0.03%

	Enzyme	Function	Form	Concentration
	Glucanases (Sheppard, 1986)	Assist filtration or lautering Provide additional sugar for fermentation	Liquid enzyme	~ 0.1% DS ~ 0.1% DS
	Cellulases (Novo Bulletin)	Assist in filtration by hydrolysing complex cell wall materials	Liquid or powder	~ 0.1% DS
	Protease (Novo Bulletin)	Provide nitrogen for yeast growth. Aid in filtration and chill proofing	Liquid or powder	~ 0.3% DS
	Diacetyl reductase	Removal of diacetyls in beer	Liquid	—
	Pectinases	*Wine* Clarification of wine. Decrease pressing time and increase the extract yield	Mainly liquid complex	~ 0.01–0.02%
	Anthocyanase	Decolorise wines	Powder or liquid	0.1–0.3%
	Amyloglucosidase	Remove haze and improve filtration	Liquid or powder	0.002% w/v
	Glucose oxidase	Remove oxygen	Powder/liquid	10–70 GOU l^{-1}
	Cellulases	*Coffee* Cellulose breakdown during drying	Liquid or powder	—
	Pectinases	Removal of gelatinous coating during fermentation	Liquid or powder	20–50 ppm
	Cellulases	*Tea* Cellulose breakdown during fermentation		—
Non-alcoholic beverages	Catalases	*Soft Drinks* Stabilisation of citrus terpenes	Powder/liquid	Combined with glucose oxidase
	Glucose oxidases	Stabilisation of citrus terpenes	Powder/liquid	20–90 GOU l^{-1}
	Pectinases	*Cocoa* Hydrolysis of pulp from beans during fermentation	Liquid	11–20 ppm

Table 6.4 *Continued*

Food group	Enzyme	Specific application	Mode of use	Dosage level or operational life
		Milk		
	Catalase	Removal of H_2O_2	Liquid or powder	—
	β-Galactosidase	Prevent grainy texture	Immobilised	Several weeks
		Stabilisation of proteins during freezing	Immolibised	Several weeks
	Protease	Stabilisation of evaporated milk	Immobilised	Several weeks
Dairy		*Cheese*		
	Proteases (Weiland, 1972)	Coagulation of casein	Powder or solution	~ 0.01–0.15%
	Lipase (Boyce, 1986)	Flavour development	Liquid or powder	~ 1% DS
		Juices		
Fruits and vegetables	Amylases	Remove starch to improve appearance and extraction	Liquid or powder	0.0005–0.002% w/v
	Cellulases	Improve extraction	Mostly liquid	0.0002–0.002% w/v
	Pectinases (Boyce, 1986)	Improve extraction	Mostly liquid	0.003–0.03%
		Aid clarification	Mostly liquid	0.01–0.02%
	Glucose oxidase	Remove oxygen	Powder/liquid	20–200 GOU1^{-1}
	Naringinase	Debittering of citrus juice	Powder	—
		Vegetables		
	Amylases	Preparation of purées and tenderisation	Mostly liquid	—
	Pectinase	Production of hydrolysates	Mostly liquid	0.02–0.03%

Food category	Enzyme	Application	Form	Usage level
Meat and other proteinaceaus foods	Proteases	*Meats and fish* Tenderisation of meat	Liquid	Variable depending on specific application and enzyme type
		Produce fish hydrolysate	Liquid	~ 2% Protein content
		Enhance fish stick-water effluent treatment	Liquid	~ 0.2%
		Removal of oil from tissues	Liquid	—
	Glucose oxidase	*Egg and egg products* Glucose removal from dried eggs	Powder/liquid	150–225 $GOU\,l^{-1}$ white 300–375 $GOU\,l^{-1}$ whole
	Lipases	Improve emulsification and whipping properties	Immobilised or powder	—
	Protease	Improve drying properties	Liquid/powder	—
Fats and oils	Pectinases	*Vegetable oil extractions* Degrade pectic substances to release oil	Liquid/powder	0.5–3% DS
	Cellulases	Hydrolyse cell wall materials	Liquid/powder	0.5–2% DS
	Lipases	*Oil Hydrolysis* To produce free fatty acids		~ 2% DS
	Esterase	*Ester synthesis* Production of terpence esters for flavour enhancement from an organic acid or alcohol	Powder (limited) Mostly immobilised systems	~ 2% DS ~ 2% E/S on batch systems
	Lipases	*Interesterification* Production of value-added triacyglycerols from less valued feed stock	Mostly immobilised systems	~ 1–5% E/S

Table 6.5   Enzymes and sources permitted for use in the United Kingdom

Enzyme	Sources
Acid proteinase (includes pepsin and chymosin)	Porcine gastric mucosa; abomasum of calf, kid or lamb; adult bovine abomasum; *Mucor miehei, M. pusillus*
α-Amylase	Porcine or bovine pancreatic tissue, *Aspergillus niger, A. oryzae, Bacillus licheniformis, B. subtilis*
Bromelain	*Ananas bracteatus, A. comosus*
Catalase	Bovine liver, *Aspergillus niger*
Cellulase	*Aspergillus niger, Trichoderma viride*
Dextranase	*Penicillium funiculosum, P. lilacinum*
Endothia carboxyl proteinase	*Endothia parasitica*
β-D-Fructofuranosidase (invertase)	*Saccaromyces cerevisiae*
β-D-Galactosidase (lactase)	*Aspergillus niger*
endo-1,3(4)-β-D-Glucanase (laminarinase)	*Aspergillus niger, Bacillus subtilis, Penicillium emersonii*
Glucose isomerase	*Bacillus coagulans*
Glucose isomerase (immobilised)	*Bacillus coagulans, Streptomyces olivaceous*
Glucose oxidase	*Aspergillus niger*
exo-1,4-α-D-Glucosidase (glucoamylase)	*Aspergillus niger*
exo-1,4-α-D-Glucosidase (immobilised)	*Aspergillus niger*
Neutral proteinase	*Aspergillus oryzae, Bacillus subtilis*
Papain/chymopapain	*Carica papaya*
Pectin esterase	*Aspergillus niger*
Pectin lyase	*Aspergillus niger*
Polygalacturonase	*Aspergillus niger*
Pullulanase	*Klebsiella aerogenese*
Serine proteinase (including trypsin)	Porcine or bovine pancreatic tissue, *Bacillus licheniformis, Streptomyces fradiae*
Triacylglycerol lipase	Edible oral forestomach tissues of the calf, kid, or lamb; porcine or bovine pancreatic tissues

From Denner (1983).

In France and Germany, enzymes are regarded as food additives, but while their use is regulated by a decree in France, they can be used without authorisation in Germany. The Danish regulation on enzymes is strict; the use of the enzyme has to be approved by the Danish National Food Institute.

In Canada enzymes are treated as food additives and are regulated accordingly. The enzymes currently approved for use in foods are presented in Table 6.6.

In the United States of America, an enzyme could be used in foods if it has 'generally recognised as safe' (GRAS) status (Table 6.7). Enzymes not on the GRAS list are considered as additives and require approval before they can be used in foods.

In Japan the regulations tend to consider commercial enzymes a synthetic

Table 6.6 Enzymes permitted for use in foods in Canada*

Enzyme	Source	Permitted in or upon
Amylase	*Bacillus subtilis, Aspergillus niger*, var.; *Aspergillus oryzae*, var.	Ale, beer, light beer, malt liquor, porter, stout
Bovine rennet	Aqueous extracts from adult bovine, sheep, and goats	Cheese, cottage cheese, cream cheese, cream cheese with named additives, cream cheese spread, cream cheese spread with named additives
Bromelain	Pineapples	Ale, beer, light beer, malt liquor, porter, stout; bread, flour, whole wheat flour, edible collagen; sausage casings, hydrolysed animal, milk, and vegetable proteins, meat cuts; meat-tenderising preparations; pumping pickle for curing beef cuts
Catalase	*Aspergillus niger, Micrococcus lysodeikticus,* bovine liver	Soft drinks, egg albumin
Cellulase	*Aspergillis niger*	Distiller's mash, liquid coffee concentrate, spice extracts, natural flavour and colour
Ficin	Fig tree latex	Same uses as bromelain except not used in bread, flour, whole wheat flour, and pumping pickles
Glucoamylase (amyloglucosi-dase maltase)	*Aspergillus niger, A. oryzae*	Ale, beer, light beer, malt liquor, porter, stout, bread, flour, whole wheat flour, chocolate syrup, distiller's mash, precooked (instant) cereals, starch used in the production of dextrins, maltose, dextrose, glucose (glucose syrup), or glucose solids, unstandardised bakery products
	Rhizopus niveus	Distiller's mash
	Rhizopus delemar	Mash destined for vinegar making
	Multiplici sporus	Brewer's mash, distiller's mash, mash for vinegar making, starch used in production of dextrins, maltose, dextrose, glucose (glucose syrup), or glucose solids (dried glucose syrup)
Glucanase	*Aspergillus niger, Bacillus subtilis*	Ale, beer, light beer, malt liquor, porter, stout, corn for degerming distiller's mash, mash destined for vinegar manufacture, unstandardised bakery products
Glucose oxidase	*Aspergillus niger*	Soft drinks, liquid whole egg, egg white, and liquid egg yolk destined for drying
Glucose isomerase	*Bacillus coagulans*, var; *Streptomyces; Olivochromogens*, var.; *Actinoplanes missouriensis*, var.; *Streptomyces olivacens*, var.	Glucose (glucose syrup) partially or completely isomerised to fructose

Table 6.6 *Continued*

Enzyme	Source	Permitted in or upon
Hemicellulase	*Bacillus subtilis*	Distiller's mash, liquid coffee concentrate, mash destined for vinegar manufacture
Invertase	*Saccharomyces*	Soft-centred and liquid-centred confections, unstandardised baking foods
Lactase	*Aspergillus niger, A. oryzae, Saccharomyces*	Lactose-reducing enzyme preparations, milk destined for use in ice cream mix
Lipase	*Aspergillus niger, A. oryzae;* edible forestomach tissue of calves, kids, or lamb; animal pancreatic tissue	Dairy-based flavouring preparations, liquid and dried egg white, Romano cheese
Lipoxidase	Soybean whey or meal	Bread, flour
Milk-coagulating enzyme	*Mucor miehei, M. pusillus, Endothia parasitica*	Cheese, cottage cheese, sour cream, Emmentaler (Swiss) cheese
Pancreatin	Pancreas of the hog or ox	Liquid and dried egg white, precooked instant cereals, starch used in the production of dextrins, maltose, dextrose, or glucose, or glucose solids
Papain	Fruit of papaya	Same as for bromelain and also for beef before slaughter and precooked instant cereals
Pectinase	*Aspergillus niger, Rhizopus oryzae*	Cider; wine; distiller's mash; juice of named fruits; natural flavour and colour extractives; skin of citrus fruits destined for jam, marmalade, and candied fruit production; vegetable stock for use in soup manufacture
Pentosanase	*Aspergillus niger*	Same as for glucanase
Pepsin	Glandular layer of porcine stomach	Ale, beer, light beer, malt liquor, porter, stout, cheese, cottage cheese, cream cheese, cream cheese (with named ingredients), cream cheese spread (with named ingredients), defatted soy flour, precooked (instant) cereals
Protease	*Aspergillus oryzae, A. niger, Bacillus subtilis*	Same as for bromelain and also dairy-based flavouring preparations, distiller's mash, industrially spray dried cheese powder, precooked (instant) cereals, unstandardised bakery foods
Rennet	Aqueous extracts from fourth stomach of calves, kids, or lambs	Same as for bovine rennet and in unstandardised milk-based dessert preparations

*Levels permitted according to Good Manufacturing Practice (GMP).

Table 6.7 Enzymes on GRAS status of the Food and Drug Administration of Health and Human Services in the United States

Enzyme	Source	Regulatory status
α-Amylase	*Aspergillus niger*	GRAS[1]
	A. oryzae	GRAS
	Rhizopus oryzae	GRAS
	Bacillus stearothermophilus	Petitioned for GRAS
	B. subtilis	GRAS (petitioned)
	B. licheniformis	184.1027
	Barley malt	GRAS
β-Amylase	Barley malt	GRAS
Bromelain	Pineapples	GRAS
Catalase	*Micrococcus lysodeikticus*	173.130
	Aspergillus niger	GRAS
	Bovine liver	GRAS
Cellulase	*Aspergillus niger*	GRAS
	Trichoderma reesei	Petitioned for GRAS
Esterase/lipase	*Mucor miehei*	173.140
	Bacillus licheniformis	184.1027
	Aspergillus niger	GRAS
	Bacillus subtilis	GRAS
Ficin	Figs	GRAS
α-Galactosidase	*Morteirella vinaceae* var. *raffinoseutilizer*	173.145
Glucoamylase or	*Aspergillus niger*	GRAS
amyloglucosidase	*A. oryzae*	GRAS
	Rhizopus oryzae	173.145
	R. niveus	173.110
Glucose isomerase	*Streptomyces rubiginosus*	184.1372
(immobilised	*Actinoplanes missouriensis*	184.1372
preparations)	*Streptomyces olivaceaus*	184.1372
	S. olivochromogenes	184.1372
	Bacillus coagulans	184.1372
	Arthrobacter globiformis	Petitioned for GRAS
	Streptomyces murinus	Petitioned for GRAS
Glucose oxidase	*Aspergillus niger*	GRAS
β-Glucanase	*Aspergillus niger*	GRAS
Invertase	*Saccharomyces cerevisiae*	GRAS
Lactase	*Aspergillus niger*	GRAS
	A. oryzae	GRAS
	Kluyveromyces fragilis	GRAS
	Candida pseudotropicalis	Petitioned for GRAS
Lipase	Calf, kid, lamb pancreatic tissue	GRAS
	Aspergillus niger	GRAS
	A. oryzae	GRAS
Milk-clotting enzyme	*Endothia parasitica*	173.150(a)(1)
	Bacillus cereus	173.150(a)(2)
	Mucor pusillus Lindt	173.150(a)(3)
	M. miehei Cooney et Emerson	173.150(a)(4)
Papain	Papaya	GRAS
Pectinase	*Aspergillus niger*	GRAS
	Rhizopus oryzae	173.130
Pepsin	Porcine and bovine stomachs	GRAS
Rennet	Ruminant fourth stomach (abomasum)	GRAS
Trypsin	Porcine, bovine pancreas	GRAS

Adapted from Denner (1983).
[1]GRAS – Generally Recognised as Safe.

and, therefore, require both approval and listing by the Food Sanitation Investigation Council.

6.6 Handling of enzyme preparations

Safe handling of chemical additives is of prime importance in the food industry. Similar precautionary measures, as used for chemicals, could conveniently be extended to enzymes. Some salient recommendations on enzymes handling by an enzyme manufacturer are summarised below (Boyce, 1986).

1. Facilities should be easy to clean. Cleaning of equipment and spills by washing, vacuum-cleaning or wet-mopping prevents the generation of dust or aerosol. Brushing, sweeping, air or high-pressure cleaning are not recommended.
2. Operations should be carried out in well-ventilated areas.
3. Proper equipment preventing dust or aerosol formation should be used.
4. Spills should be prevented as much as possible. If spills occur, they should be cleaned as described in 1 above.
5. There should be a plan in place for handling enzymes.
6. Personnel should be properly educated on enzyme safety.
7. Protective gear, to prevent direct contact of the enzyme with parts of the body, are recommended to be worn.
8. General personal hygiene and carefulness are prerequisities for safe handling of enzymes.

References

AACC (1983) Nephelometric alpha amylase analysis. In *AACC Approved Methods* #22-07. American Association of Cereal Chemists Inc., St Paul, MN.

AACC (1983) Diastatic activity of flour with amylograph. In *AACC Approved Methods* #22-10. American Association of Cereal Chemists Inc., St Paul, MN.

Adler-Nissen, J. L. (1977) Enzymatic hydrolysis of food proteins. *Process Biochem.* **12** (6), 18.

Adler-Nissen, J. L. (1979) Determination of the degree of hydrolysis of food protein hydrolysates by trinitrobenzenesulfonic acid. *J. Agric. Food Chem.* **27** (6), 1256–1262.

AOCS (1981) Official American Oil Chemists Society Method #Ce 1-62.

AOCS Official Methods #Cd 3a-63, #Cd 11-57.

Ardeshir, H. (1987) Enzymes in food processing. *Nutrition and Food Science* (July/August) **107**, 12–14.

ASBC (1958) Iodine reaction. In *Methods of Analysis*. American Society of Brewing Chemists Inc., St Paul, MN.

Barman, T.E. (1969) *Enzyme Handbook*, Vols 1 and 2. Springer Verlag Chemie, New York.

Bates, F. J. (1942) *Polarimetry, Saccharimetry of Sugars*, NBS Circular #440. National Bureau of Standards, Washington, DC.

Bergmeyer, H. U. 1974. *Enzymatic Methods in Analysis*. Springer-Verlag, Düsseldorf.

Bigelis, R. and Lasure, L. L. (1987) Fungal enzymes and primary metabolites used in food processing. In *Food and Beverage Mycology* (ed. Beuchat, L. R.). Van Nostrand Reinhold, New York.

Birch, G. G., Blakebrough, N. and Parker, K. J. (1981) *Enzymes and Food Processing*. Applied Science Publishers, London.

Bitman, J. and Wood, D. L. (1981) Quantitative densitometry *in situ* of lipid separated by thin layer chromatography. *J. Liq. Chromatogr.* **4**, 1023–1034.

Boyce, C. O. L. (1986) *Novo's Handbook of Practical Biotechnology*. Novo Industries, Bugsvaerd, Denmark.

Colwick, S. P. and Kaplan, N. O. (1955) *Methods in Enzymology*. Academic Press, New York.

Dealy, J. M. (1984) Viscometers for online measurements and control. *Chem. Eng.* 1 October, 62–70.

Denner, W. H. B. (1983) The legislative aspects of the use of industrial enzymes in the manufacture of food and food ingredients. In *Industrial Enzymology: The Application of Enzymes in Industry* (eds Godfrey, T. and Reichett, J.). The Nature Press, New York.

Dole, V. P. and Mienertz, H. (1960) Microdetermination of long-chain fatty acids in plasma and tissue. *J. Biol. Chem.* **235**, 2597–2599.

Dupuy, P. (1982) *Uses of Enzymes in Food Technology*. Proc. of Internation. Symposium. Technique of Documentation, Lavoisier, Paris.

Engle, C. E. and Olinger, P. M. (1979) HPLC determination of saccharides in corn syrup. *J. Assoc. Off. Anal. Chem.* **62** (3), 527.

Fitton, M. G. (1979) Rapid determination of dextrose equivalent by cryoscopy. *Stärke.* **11**, 381–384.

Folch, J., Lees, M. and Stanley, G. H. S. (1957) A simple method of isolation and purification of total lipid from animal tissues. *J. Biol. Chem.* **226**, 497–509.

Fullmer, C. S. and Wasserman, R. H. (1979) Analytical peptide mapping by high performance liquid chromatography. *J. Biol. Chem.* **254** (15), 7208–7212.

Ghildyal, N. P., Ramakrishna, S. V., Losane, B. K. and Ahmed, S. Y. (1979) Immobilized enzymes: application in food industry. *Proceedings First National Immobilized Enzyme Engineering, Calcutta*, pp. 139–150.

Godfrey, T. (1983) Immobilized enzymes. In *Industrial Enzymology: The Application of Enzymes in Industry* (eds Godfrey, T. and Reichelt, J.). The Nature Press, New York.

Godfrey, T. and Reichelt, J. (eds) (1983) *Industrial Enzymology: The Application of Enzymes in Industry*. The Nature Press, New York.

Gunther, R. W. (1979) Chemistry and characteristics of enzyme modified whipping proteins. *J. Am. Oil Chem. Soc.* **56**, 345–459.

Jacobsen, F. and Rasmussen, O. L. (1984) Energy savings through enzymatic treament of stickwater in the fish meal industry. *Process Biochem.* **19** (5), 165–169.

Jensen, R. G. (1981) Improved separation of triglycerides at low temperature by reversed phase liquid chromatography. *J. Chrom.* **204**, 407–411.

Jensen, R. G. (1983) Detection and determination of lipases (acylglycerol hyrolysis) activity from various sources. *Lipids* **18**, 650–657.

Kilara, A. (1982) Enzymes and their uses in the processed apple industry: a review. *Process Biochem.* **17** (4), 35–44.

Klacik, M. A. (1988) Enzymes in food processing. *Chemical Engineering Progress* **84** (5), 25–29.

Kupier, J., Roels, J. A. and Zwidweg, M. H. V. (1978) Flow-thru viscometer for use in the automated determination of hydrolytic enzyme activities. *Anal. Biochem.* **90**, 192–203.

Law, B. A. and Wigmore, A. (1982) Accelerated cheese ripening with food grade proteinases. *J. Dairy Res.* **49**, 137–146.

Leiva, M. L. and Gekas, V. (1988) Application of immobilized enzymes in food processing. In *Developments in Food Microbiology – 4*. Elsevier Applied Science, London.

Madsen, G. B., Norman, B. E. and Slott, S. (1973) A new heat stable bacterial amylase and its use in high temperature liquefaction. *Stärke* **25**, 304–308.

Moore, S. and Stein, W. H. (1948) Photometric ninhydrin method for use in the chromatography of amino acids. *J. Biol. Chem.* **176**, 367–388.

Moskowitz, G. J., Cassaigne, R., West, R. J., Shen, T. and Feldman, L. I. (1977) Hydrolysis of animal fat and vegetable oil with *Mucor miehei* esterase. *J. Agric. Food Chem.* **25**, 1146–1150.

NAS/NRC (1981) *Food Chemical Codex* (3rd edn) National Academy of Sciences, National Research Council, Food and Nutrition Board Committee on Codex Specification. National Academy Press, Washington, DC.

Neilson, J. M. and Griffin, E. G. (1916) Adsorption of invertase. *J. Am. Chem. Soc.* **38**, 1109–1115.

Nijpels, H. H., Evers, P. H., Noval, G. and Ramet, J. P. (1980) Application of cryoscopy for measurement of enzymatic hydrolysis of lactose. *J. Food Sci.* **45**, 1684–1687.

Novo Industries Method No. F-840289. *Isolation of Pure Triglyceride from Fat Samples.*

Okada, S., Inoue, M. and Fukumoto, J. (1968) Purification and some properties of a thermostable anthocyanase. *Kaguku To Kugyo* **46** (6), 303–308.

Olsen, H. S. (1984) Enzyme technology in modern food processing. *Biotechnology and Farming: The Energetic Cultivations.* Forli, Italy.

Olsen, H. S. and Christensen, F. M. (1987) Novel uses of enzymes in food processing. *7th World Congress of Food Science and Technology, Singapore.*

Ory, R. L. and Angelo, A. J. (1977) *Enzymes in Food and Beverage Processing.* ACS Symposium Series 47. American Chemical Society. Washington, DC.

Payne-Wahl, K., Spencer, G. F., Plattner, R. D. and Butterfield, R. O. (1981) High performance liquid chromatographic method of quantitation of free acids, mono-, di-, and triglycerides using infrared detector. *J. Chromat.* **209**, 61–66.

Poulsen, P. B. (1984) Current applications of immobilized enzymes for manufacturing purposes. *Biotechnol. Genet. Eng. Reviews* **1**, 121–140.

Reed, G. and Underkofler, L. A. (1966) *Enzymes in Food Processing.* Academic Press, New York.

Regnier, F. E. (1983) High performance liquid chromatography of biopolymers. *Science* **222**, 245–252.

Sampugna, J., Quinn, J. G., Pitas, R. E., Carperter, D. L. and Jensen, R. G. (1967) Digestion of butyrate glycerides by pancreatic lipase. *Lipids* **2**, 397–402.

Schwimmer, S. (1981) *Sourcebook of Food Enzymology.* AVI, Westport, Conn.

Scott, D. (1953) Glucose conversion in preparation of albumin solids by glucose oxidase–catalase reaction system. *J. Agr. Food Chem.* **1**, 727–730.

Sheppard, G. (1986) The production and uses of microbial enzymes in food processing. *Progress in Industrial Microbiology* **23**, 237–283.

Shulka, V. K. S., Nielsen, W. S. and Batsberg, (1981) A simple and direct procedure for the evaluation of triglyceride composition of cocoa butters by HPLC. A comparison with existing TLC-GC methods. *Fette Seifen Anstrichmittel.* **7**, 274–278.

Sonntag, N. O. V. (1982) Glycerolysis of fats and methyl esters status, review and critique. *J. Am. Oil Chem. Soc.* **59**, 795A–802A.

Strobel, R. J., Ciavarelli, L. M., Starnes, R. L. and Lanzilotta, R. P. (1983) Biocatalytic synthesis of esters using dried *Rhizopus arrhizus mycelium* as a source of enzyme. Abstract of American Society for Microbiology Meeting, New Orleans, LA.

Swaisgood, H. E. (1985) Immobilization of enzymes and some applications in the food industry. In *Biotechnology Series 5.* Benjamin/Cummings, Menlo Park, California.

Taylor, M. J. and Richardson, T. (1979) Application of microbial enzymes in food systems and in biotechnology. *Advances in Applied Microbiology* **25**, 7–35.

Underkofler, L. A., Baron, R. R. and Aldrich, F. L. (1960) Methods of assay for microbial enzymes. In *Developments in Industrial Microbiology*, **Vol. 2**. Plenum Press, New York, pp. 171–182.

Wieland, H. (1972) *Enzymes in Food Processing and Products.* Noyes Data Corporation, Park Ridge, New Jersey.

Whitaker, J. R. (1974) *Food Related Enzymes. Advances in Chemistry Series 136.* American Chemical Society, Washington, DC.

Woo, A. H. and Lindsay, R. C. (1982) Rapid and quantitative analysis of free fatty acid in cheddar cheese. *J. Dairy Sci.* **65**, 1102–1109.

7 Nutritive additives

D. PEACE and L. DOLFINI

7.1 Vitamins

7.1.1 *Fat-soluble vitamins*

The fat-soluble vitamins are A, D, E and K. To fortify foods containing a significant proportion of fat, oil-based dilutions of the fat-soluble vitamins are available. In order to add fat-soluble vitamins to aqueous foods, it may be possible to use products which are compounded with such water-soluble substances as gelatin, sucrose and acacia. These water-dispersible forms of the fat-soluble vitamins contain either the natural antioxidants, Vitamin E, ascorbyl palmitate, ascorbic acid, or synthetic compounds BHA, BHT. A synopsis of fat-based and water-dispersible forms of the fat-soluble vitamins is found in Table 7.1.

Of the fat-soluble vitamins, Vitamin A is the most labile, being sensitive to oxygen, light and heat. β-Carotene (pro-Vitamin A), has similar characteristics. Vitamin E, when used as a nutrient source, should be purchased in its esterified form (Vitamin E acetate). The alcohol form is a useful antioxidant. Vitamin K is seldom added to foods other than infant formulae and meal replacements.

Fat-soluble vitamins can be stored in a cool room (or refrigerator) in their unopened containers for six months to a year. The oily forms of Vitamin A have better stability than the dry forms. After six months, Vitamin A products should be assayed to confirm potency. Vitamins in oil-based products may crystallise in the cold; therefore, they should be warmed to room temperature prior to using.

7.1.2 *Water-soluble vitamins*

The water-soluble vitamins include Vitamin C and the Vitamin B group – thiamin, riboflavin, niacin/niacinamide, pyridoxine, cyanocobalamin, folic acid, pantothenic acid and biotin. The most commonly used forms of the water-soluble vitamins are given in Table 7.2. These vitamins are quite stable when kept dry in their unopened containers. A shelf-life of at least one year can be expected.

Table 7.1 Fat-soluble vitamins

Vitamin	Chemical name	Description	Potency	Usage	Usage considerations
Vitamin A	Retinol				
	Retinyl acetate Retinyl palmitate	Oily blends: yellow oil, stabilised with tocopherol or BHA/BHT	$1-2 \times 10^6$ IU g$^{-1(1)}$ Custom blends	Fortification of oils and fats, liquid milk, milk powder, infant formulae, breakfast cereals	Soluble in oil and fat; use minimal agitation, protect from light, oxygen and extreme heat; may crystallise on storage
					Homogenising in a small amount of liquid is necessary prior to adding to milk and infant formulae
		Dry powders: light yellow, fine granular powder, stabilised with tocopherol, ascorbic acid, ascorbyl palmitate or BHA/BHT	$0.25-0.5 \times 10^6$ IU g^{-1}	Fortification of: milk powder, infant formulae, flour, breakfast cereals, soups, milk modifiers	Water dispersible; use minimal agitation, protect from light, oxygen and extreme heat
Vitamin D$_2$	Ergocalciferol	Crystalline white powder			Mostly substituted by Vitamin D$_3$

		Form	Concentration	Use	Solubility
Vitamin D₃ [2]	Cholecalciferol	Crystalline white powder	40×10^6 IU g^{-1}	Frequently diluted for use as stated below	Soluble in oil and fat; homogenising in a small amount of liquid is necessary prior to adding to milk and infant formulae
		Oily blends: clear, colourless to slightly yellowish oil, stabilised with tocopherol	1.0×10^6 IU g^{-1}	Fortification of: liquid milk, milk powder, oils and fats, breakfast cereals and other cereal products	
		Dry powders: off-white to yellowish, fine granular powder	$0.1\text{–}0.85 \times 10^6$ IU g^{-1}	As above	Water dispersible
Vitamin E	Tocopherol				
	dl-α-Tocopherol	Yellow, viscous oil	1000 IU g^{-1}	As antioxidant in oils, fats, fat based products, sausages	Soluble in fat and oil
	dl-α-Tocopheryl-acetate	Oily blends: slightly yellow viscous oil		Fortification of infant formulae, sugar and cocoa confectionery, oils and fats, fruit drinks, flour, liquid milk and milk powder, breakfast cereals	Soluble in fat and oil; homogenising in a small amount of liquid is necessary prior to adding to milk, infant formulae and drinks
		Dry powders: slight yellowish free-flowing powder	$0.25\text{–}0.50 \times 10^6$ IU g^{-1}	As above	Water dispersible

Table 7.1 *Continued*

Vitamin	Chemical name	Description	Potency	Usage	Usage considerations
Natural	tocopherols: mixture of d-α-, d-β-, d-γ- and d-δ-tocopherol forms	Oily blends: red to reddish-brown slightly viscous liquid	50–70%	As antioxidant in: animal fat, margarine, sausages, poultry products, shrimps (breaded), pasta, bakery products, snack foods, confectionery products, sauces, dehydrated vegetables	In some cases unsaturated vegetable oils do not benefit from adding a supplement of tocopherols because of the optimal level of tocopherols naturally present in oils after refining and processing (0.05–0.1%)
		Dry blends: off-white to pale yellow; not free flowing	30%	As above (without fat and margarine)	
Vitamin K$_1$	Phytonadione or phylloquinone	Oily blends: clear yellow to amber, viscous oil stabilised with tocopherol	100%	Fortification of: infant formulae, liquid milk, oils and fats, dietetic products	Soluble in fat and oil; homogenising in a small amount of liquid is necessary prior to adding to milk and infant formulae
		Dry powders: off-white to yellow, free-flowing powder	10–50 mg g^{-1}	As above	Water dispersible

Provitamin A	β-Carotene	Oily suspensions: brick-red viscous oil stabilised with tocopherol, ascorbyl palmitate	20–30%	Coloration and fortification of oils and fats, dressings, butter, ice cream, confectionery, fruit drinks	Soluble in warm fat or oil (50°C) Colour range: yellow to orange
		Emulsions: oil in water, stabilised with tocopherol or BHA/BHT	0.5–5%	Coloration and fortification of fruit drinks, dairy products, pasta, bakery products, snacks, confectionery, dressings	Water dispersible Colour range: yellow to orange
		Dry powders: red-brown, fine granular powder	2.4–10%	Coloration and fortification of dairy products, pasta, snacks, bakery products, confectionery, beverages, soups, sauces, dressings	Dispersible in hot water (60°C) make stock solutions Colour range: orange
		Fine orange powder	1%	Coloration and fortification of instant products, confectionery, dairy products, pasta, snacks, bakery products	Cold water dispersible Colour range: yellow

[1]IU g^{-1} = International Units per gram.
[2]Combinations of Vitamins A and D$_3$ are available in dry and oily forms.

Table 7.2 Water-soluble vitamins

Vitamin	Chemical name	Description	Potency	Usage	Usage considerations
Vitamin B_1	Thiamin		100%	Not used in its pure form	
	Thiamin hydrochloride	White or almost white powder	89.3%	Fortification of flour, breakfast cereals, infant formulae, soups, sugar and cocoa confectionery, oils and fats, milk drinks, pasta, meal replacements	Thiamin hydrochloride is more water soluble than thiamin mononitrate
	Thiamin mononitrate		92.0%	Used as reaction flavour component in flavouring for soups and sauces (meat flavour)	
		Coated form	33.3%	Used in dry products to mask taste	
Vitamin B_2	Riboflavin	Yellow to orange-yellow powder	100%	Fortification of flour, breakfast cereals, sugar and cocoa confectionery, soups, infant formulae, fruit drinks, oils and fats, desserts, milk drinks, meal replacements	Riboflavin-5'-phosphate is more water soluble than pure riboflavin and is particularly suitable for coloration or instant products
	Riboflavin-5'-phosphate sodium salt	Yellow to orange powder	78.7%	Coloration of ice creams, desserts, instant beverages, sauces, soups, confectionery products, pasta	
		Coated forms	25-33.3%		Protect solutions from light

Vitamin B$_6$	Pyridoxine		100%	Not used in its pure form	
	Pyridoxine-hydrochloride	White or almost white crystalline powder	82.0%	Fortification of breakfast cereals, sugar and cocoa confectionery, infant formulae, fruit drinks, flour, cereal products, oils and fats, milk drinks, meal replacements	
		Coated form	33.3%		
Niacin	Nicotinic acid	White crystalline powder	100%	Fortification of flour, breakfast cereals, infant formulae, fruit drinks, sugar and cocoa confectionery, pasta, meal replacements	Niacin may cause vasodilation, niacinamide does not
	Niacinamide (nicotinamide)			Colour stabilisation in meat products (if permitted)	Addition level: 50–100 ppm
		Coated form	33.3%		Niacinamide may cake
Pantothenic acid	Pantothenic acid		100%	Not used in its pure form	
	Calcium-D-pantothenate	White powder	92.0%	Fortification of infant formulae, breakfast cereals, fruit drinks, sugar and cocoa confectionery, milk drinks, meal replacements	
Vitamin B$_{12}$	Cyanocobalamin	Dark red crystalline powder	100%	Fortification of infant formulae, meal replacements and substitute foods	Hygroscopic; serial dilution required
		Fine pink powder diluted with sugar starch or dicalcium phosphate	0.1–1.0% dilutions	As above	Good distribution

Table 7.2 *Continued*

Vitamin	Chemical name	Description	Potency	Usage	Usage considerations
Folic acid	Pteroylglutamic acid	Yellow to orange crystalline powder	90%	Fortification of infant formulae, breakfast cereals, fruit drinks, sugar and cocoa confectionery, milk drinks, meal replacements	Serial dilution may be required
Biotin	Biotin	White crystalline powder	100%	Fortification of infant formulae, meal replacements and substitute foods	Serial dilution required
Vitamin H		Dilution with dicalcium phosphate	1% trituration	As above	Good distribution
Vitamin C	Ascorbic acid	White to slightly yellow crystalline powder	100%	Fortification of infant formulae, meal replacements, fruit juices and drinks, cereal products, dairy products, sugar and cocoa confectionery	Protect from oxygen and extreme heat
		Various granulations available		As antioxidant in fruit juices, soft drinks, beer, wine, canned fruit and vegetables, potato products, dairy products	
				As flour and bread improver	50–200 ppm
				As a curing agent and nitrosamine inhibitor in sausage and other comminuted meats	300–500 ppm
		Fat-coated ethyl cellulose coated	96% 97.5%	Fortification of milk modifiers, cocoa drink powders, bakery products	
				As antioxidant in processed potatoes, sausages	Suitable for dry products

Sodium ascorbate	White to slightly yellowish powder, various granulations available	88%	As a curing agent and nitrosamine inhibitor in cured meats; fortification of dairy products	300–500 ppm
Calcium ascorbate	White powder	83%	Fortification of breakfast cereals, low sodium dietetic products	Bitter taste
Ascorbyl palmitate	White to yellowish powder	43%	As antioxidant in oils, fats, fat-based products, uncured frozen sausages, processed potatoes, extruded cereals	Solubility in fat and oil: $30 \text{ mg } (100 \text{ ml})^{-1}$

7.1.3 *Vitamins in food processing*

Added vitamins are generally as stable as those that occur in a product naturally. In order to estimate the extra amount of vitamins required to account for losses during processing and shelf-life of the product, the food scientist must consider the nature of the product, the methods used in processing, the packaging and the conditions under which the product will be stored.

Those factors which affect vitamin stability are listed in Table 7.3. Table 7.4 provides a list of overages above label claim generally required to maintain vitamin stability for one year in a bottled, pasteurised, multi-vitaminised fruit drink. Some other processes (e.g. extrusion or baking) may require higher overages; aseptic packaging, freezing and dry blends will probably require less.

All vitamin-fortified products should undergo stability testing in the development stage. The product should be stored under conditions which simulate or exceed conditions of normal use. For example, under accelerated testing conditions, the product is kept at 35 or 45°C and 45% humidity for several weeks. Vitamin assays should be performed at time 0, at 1 week, 2 weeks and at monthly intervals for three months. Along with potency testing, the scientist should perform sensory evaluation.

Vitamin suppliers will often provide analytical services for customers developing new products.

7.2 Minerals

Minerals are often added to foodstuffs, together with vitamins. When evaluating minerals for food enrichment, the following criteria should be considered: moisture content/hygroscopicity, particle size compatibility, solubility, pH effect, taste/texture, odour, colour, interaction with vitamins or other components, ease of mixing, bioavailability, safety, purity/potency, relative cost and application needs.

Examples of mineral additives are listed in Table 7.5.

7.2.1 *Minerals in food processing*

Little or no loss of minerals occurs during processing, so no significant overages are required. However, adding minerals to food may occasionally cause taste, colour, odour or stability problems. If the food product is dry (e.g. instant soups or sauces), few stability problems will occur. However, in such high-moisture foods as baked goods, cereals with fruits, and beverages, minerals may severely affect vitamin and, occasionally, lipid stability. New

Table 7.3 Factors affecting the stability of vitamins added to foods

Vitamin	Air oxidation	Light exposure	Reducing agents	Minerals	Optimum pH	Heat stable
Vitamin A	Yes	Yes	No	Yes	>6	Fair
β-Carotene	Yes	Yes	No	Yes	Not critical	Fair
Vitamin D	Yes	Yes	No	No	Not critical	Fair
Vitamin E (ester)	No	No	No	No	Not critical	Yes
Vitamin K	No	Yes	Yes	No	4–7	Yes
Vitamin C	Yes	Yes	No	Yes	5–7	No
Folic acid	Yes	Yes	Yes	No	6–9	No
Thiamin	No	No	Yes	Sulphite	3–4.5	No
Riboflavin	No	Yes	Yes	No	Not critical	Yes
Niacinamide	No	No	No	No	Not critical	Yes
Pyridoxine	No	No	No	Yes	Not critical	Fair
Vitamin B$_{12}$	Yes	No	Yes	No	4–5	Fair
Biotin	No	No	No	No	Not critical	Yes
Pantothenate	No	No	No	No	5–7	Fair

Table 7.4 Overage of vitamins in fruit-juice drinks at one year storage[1]

Vitamin	% over label claim
B$_1$	40
B$_2$	25
B$_6$	25
B$_{12}$	100
Biotin	25
Calcium pantothenate	100
Folic acid	120
Niacin	10
E	10
C	30
β-Carotene	40

[1] Ambient temperature, protected from direct light.

developments in the technology of coatings applied to minerals and vit-amins may prove beneficial in solving some taste and stability problems. However, these coatings are generally ineffective in liquid environments. Coatings include high-melting-point fats, ethylcellulose and synthetic resins.

Mineral salts often contain a low percentage of the element, requiring the use of large quantities of the salt. Other problems include poor solubility and unacceptable colours.

7.3 Vitamin and mineral premixes

Moisture content and hygroscopicity are the most important criteria to safeguard vitamin and mineral premixes. As a rule, all ingredients should contain a maximum of 5% moisture. Little or no loss in ingredient potency occurs in premixes kept in a cool, dry place and in their sealed original containers. Vitamin A potency should be checked after six months.

7.4 Government regulations

In most countries, the addition of nutrients to food is controlled by government agencies. The food chemist should be aware of the regulations of the country to which the food is destined for sale.

The Codex Alimentarius Commission of the FAO/WHO have adopted the following basic principles for the addition of nutrients to foods:

1. Restoration – the addition to a food of essential nutrients in the amounts lost during processing, storage and handling.
2. Nutritional equivalence of a substitute food – the addition to a new food of nutrients found in the traditional food.

Table 7.5 Minerals for addition to foods (anh – Anhydrous; hy – Hydrous)

Mineral	Description	% Mineral[1]	Usage considerations			
			Water soluble	Taste	Bioavailability[2]	Other
Ferrous sulphate anhydrous	White to yellow powder	36.8% anh	Very	Metallic	Excellent	Affects Vitamin C stability
Ferrous fumerate	Reddish-brown granular powder	32.9% anh	Moderate	Slight	Excellent	Affects Vitamin C stability
Ferric orthophosphate dihydrous	Greyish-white powder	29.9%	Not	Tasteless	Poor	Used when whiteness and inertness are priorities
Reduced iron/electrolytic iron	Grey-black powder	100.0%	Not	Metallic	Good	Smaller particle sizes are more bioavailable
Calcium carbonate	White powder	40.0% anh	Not	Tasteless to chalky	Fair	Incompatible with acids
Calcium glycerophosphate	Fine white powder	19.1% anh	Moderate	Tasteless	Unknown	Hygroscopic Slightly alkaline
Calcium lactate pentahydrate	White effervescent powder	18.4% anh 13.0% hy	Very (hot water)	Tasteless	Good	Neutral Hygroscopic
Calcium phosphate tribasic	White powder	38.8% anh	Not	Tasteless	Fair	Affects B_1 stability Alkaline
Calcium phosphate dibasic	White powder	30.0% anh	Not	Tasteless	Fair	Phosphorus source Affects B_1 stability Alkaline
Magnesium oxide (heavy)	White powder	60.3% anh	Moderate	Chalky	Good	Hygroscopic, soluble in acids Slightly alkaline
Magnesium carbonate hydroxide	White powder	Varies (25-30%)	Slightly effervescent	Tasteless to chalky	Exellent	Contains H_2O

Table 7.5 *Continued*

Mineral	Description	% Mineral[1]	Water soluble	Taste	Bioavailability[2]	Other
					Usage considerations	
Zinc sulphate monohydrate	White, effervescent powder	40.5% anh 36.4% hy	Very	Astringent	Good	Heptahydrate form cakes
Cupric gluconate monohydrate	Blue-green crystalline powder	14.0% anh 13.5% hy	Very	Astringent	Good	Affects Vitamin C stability

[1] '% Mineral' does not take into consideration impurites or moisture which may be present in commercial products. Check supplier's Certificate of Analysis.
[2] Bioavailability refers to the pure compound. When combined with other food components, bioavailability may be altered.

3. Fortification – the addition of one or more essential nutrients to a food over and above the levels normally contained in the food, for the purpose of correcting a deficiency in the population.
4. Ensuring the appropriate nutrient composition of a special purpose food – foods such as meal replacements and foods for special dietary use.

In the United States, food (other than standardised foods) can be fortified to levels reflecting a percentage of a standard (Recommended Daily Allowance – RDA). In other countries (for example, Canada), the specific foods which can be fortified together with the fortification levels are specified in the Food and Drug Regulations.

Nutrition labelling, a system of reporting the nutritional content of a food on its label, is widely used throughout the world. Systems vary from country to country, but usually they are based on expressing nutrient content as a percentage of a standard. Three such standards are found in Table 7.6. Two examples of nutrition labelling are found in Table 7.7. In the US, nutrition labelling is mandatory if a claim is made or if nutrients are added to foods. The Canadian system is voluntary. The addition of nutrients does not require that all nutrients be declared. However, levels of added nutrients must be clearly stated on the label as a percentage of the RDI. If a vitamin

Table 7.6 The United States, Canadian and proposed European Community recommended daily allowances for vitamins and minerals for use with nutrition labelling

Nutrient	US RDA[1]	Canadian RDI[2]	EC RDA[4]
Vitamin A	5000 IU	1000 RE	1000 μg
Vitamin D	400 IU	5 μg	5 μg
Vitamin E	30 IU	10 mg[3]	10 mg
Vitamin C	60 mg	60 mg	60 mg
Thiamin	1.5 mg	1.3 mg	1.4 mg
Riboflavin	1.7 mg	1.6 mg	1.6 mg
Niacin	20 mg	23 mg	18 mg
Vitamin B_6	2 mg	1.8 mg	2 mg
Folacin	400 μg	220 μg	400 μg
Vitamin B_{12}	6 μg	2 μg	3 μg
Biotin	—	—	0.15 mg
Pantothenic acid	—	7 mg	6 mg
Calcium	1000 mg	1100 mg	800 mg
Iron	18 mg	14 mg	12 mg
Phosphorus	1000 mg	1100 mg	800 mg
Iodine	150 μg	160 μg	150 μg
Magnesium	400 mg	250 mg	300 mg
Zinc	15 mg	9 mg	15 mg

[1] For adults and children 4 years or above.
[2] For adults and children 2 years or above.
[3] d-α-Tocopherol equivalents.
[4] Proposed: Off. J. Eur. Community, No. C282, November 1988.

Table 7.7 Examples of United States, Canadian and UK nutrition labels for one slice of processed cheese

US Nutrition Label

Nutrition information per serving	
Serving size	1 oz
Serving per pkg	8
Calories	80
Protein	5 g
Carbohydrate	2 g
Fat	6 g
Sodium	430 mg
Percentage of US Recommended Daily Allowance (US RDA)	
Protein	10
Vitamin A	4
Riboflavin	8
Calcium	15
Contains less than 2% of US RDA of Vitamin C, thiamin, niacin and iron.	

Canadian Nutrition Label

Nutrition information per 30 g serving (1 slice)	
Energy	80 cal/340 kJ
Protein	4.5 g
Fat	5.5 g
Carbohydrate	1.5 g
Percentage of Recommended Daily Intake	
Vitamin A	13
Thiamin	2
Riboflavin	8
Niacin	1
Calcium	13

UK Nutrition Label

Nutrition information	per 100 g	per slice
Energy	1250 kJ/300 kcal	250 kJ/60 kcal
Protein	20 g	4.0 g
Carbohydrate	1 g	0.2 g
Fat	24 g	4.8 g
Calcium	700 mg	140 mg

is added solely for its technical functionality (e.g. sodium ascorbate as a meat curing accelerator) nutritional labelling is not required.

A list of foods to which nutrients are commonly added is found in Table 7.8. Table 7.9 includes conversion factors for some of the new units of potency for the vitamins.

Table 7.8 Foods to which nutrients are commonly added

Food	Nutrients	Levels per serving[1]
Fruit juices	Vitamin C	1 RDA
Vegetable juices	Vitamin C	1 RDA
Fruit drinks	Vitamin C	1 RDA
	All vitamins	$\frac{1}{4}$ RDA
	Iron	Equivalent to juice
	Calcium	Equivalent to juice
	Potassium	Equivalent to juice
Dehydrated fruits and vegetables	Vitamin C	Restoration
Meal replacement foods	All nutrients	$\frac{1}{3}$ RDA
Margarine	Vitamin A	Equivalent to butter
	Vitamin D	$\frac{1}{4}$ RDA
	Vitamin E	$\frac{1}{4}$ RDA
Milk (fat altered)	Vitamin A	Restoration
	Vitamin D	$\frac{1}{4}$ RDA
Cereals	All vitamins and minerals	Restoration or $\frac{1}{4}$ to 1 RDA
Pasta	All vitamins	As above
White flour/bread	B vitamins and iron	Restoration
Rice	As above	As above
Snack foods	All nutrients	$\frac{1}{8}$–$\frac{1}{4}$ RDA

[1]Typical label claims. Actual amounts will vary according to government regulations of a country.

Table 7.9 Commonly used conversion factors for vitamin activity

Vitamin A activity

Retinol equivalents (RE) = 1 μg retinol
= 6 μg β-carotene
= 12 μg other pro-Vitamin A carotenoids
= 3.33 IU vitamin activity from retinol
= 10 IU vitamin activity from β-carotene
= 1.147 μg retinyl acetate
= 1.546 μg retinyl palmitate

Vitamin D activity

1 μg = 40 IU

Vitamin E activity

1 mg dl-α-tocopheryl acetate = 1 IU
1 mg dl-α-tocopheryl acetate = 0.67-α-tocopherol equivalents (mg)

Niacin equivalents

NE = 1 mg niacin or niacinamide
= 60 mg tryptophan
(Factors fo tryptophan content: 1.5% egg protein; 1.3% all other animal protein; 1.1% other proteins)

Further reading

Borenstein, B. and Lachance, P. A. (1988) Effects of processing and preparation on the nutritive value of foods. In *Modern Nutrition in Health and Disease* (7th edn). Lea and Febiger, Philadelphia, pp. 672–684.

Clydesdale, F. (1983) Physicochemical determinants of non-bioavailability. *Food Technology* (October), 133–144.

Hartman, G. H. (1975) Technological problems in fortification with minerals. In *Fortification of Foods: Proceedings of a Workshop*. National Academy of Sciences, Washington DC, pp. 8–18.

IFT Nutrition Division (1988) Symposium: Bioavailability of vitamins in food. *Food Technology* (October), 192–221.

Machlin, L. J. (ed.) (1984) *Handbook of Vitamins*. Marcel Dekker, New York.

Mertz, W. (1977) Fortification of foods with vitamins and minerals. *Ann. N.Y. Academy of Sciences* **300**, 151–160.

Paulus, K. (1989) Vitamin degradation during food processing and how to prevent it. In *Nutritional Impact of Food Processing*. Bibl. Nutr. Dieta, Basel Karger, pp. 173–187.

8 Emulsifiers

B. S. KAMEL

8.1 Introduction

Food product development has come a long way and the food industries have long since been able to provide the consuming public with products that have been well preserved and are safe. Present-day consumers, at their convenience and discretion, will be able to purchase these products at any time they desire. These developments have not been easy to achieve and in many cases have required special processing or preservation techniques using various food additives which permit foods to be of high quality, convenient, as well as having a longer shelf-life, easier storage conditions and to be appealing to sight, touch, taste and smell. Food emulsifiers form a category of additives that have been increasingly used by the food industry for providing these qualities.

The first food emulsifiers utilised in the food industry were natural substances such as gums, polysaccharides, saponins, lecithin, lipoproteins, liquorice, bile salts and wool wax. While these materials were major break-throughs in their day, they were incapable of meeting the diverse require-ments of an industry endeavouring to satisfy growing demands for sophis-ticated and highly formulated foods. Although some of these products are still being used, the synthetic food emulsifiers which are made by chemical processes are more commonly utilised.

There has been a tremendous amount of development in the science and technology of food emulsifiers during the first half of the twentieth century. Although the use of mono- and diglycerides in the margarine industry was reported as early as 1921, it was not until the later 1930s that the production of food emulsifiers was established industrially to any significant extent. Over the years, the impact of surfactants, from the viewpoint of their fundamental applications in our day-to-day existence, is overwhelming. These branches of chemistry and chemical technology permeate our lives in very diverse ways, as shown in Table 8.1. In the USA, a total of 118 million kilograms are used annually of all types of food grade emulsifiers. Mono- and diglycerides form the largest single type used at 85 million kilograms, which represents approximately 70% of total emulsifiers used in different

Table 8.1 Major industrial uses of surfactants

Adhesives	Leather processing
Agricultural	Metal working
Asphalt	Mould release agents
Building materials	Oil well drilling
Cement additives	Ore flotation
Coal fluidisation	Paints
Coating and levelling additives	Paper manufacture
Coolants and lubricants	Petroleum recovery
Cosmetics	Pharmaceuticals
Dry cleaning fluids	Photographic products
Electroplating	Plastic additives
Emulsion polymerisation	Surface preparations
Foods and beverages	Textile finishes
Household cleaning and laundering	Waterproofing
Industrial cleaning	

food products. With increased demand for convenience foods, the projected increase in emulsifier usage is approximately 2% per annum.

The terms *surfactant, surface-active agents, emulsifiers,* and *emulsifying agents* are all encountered in the literature and are used interchangeably, although this may not reflect exactly what the terms mean. The term *emulsifier* or *emulsifying agent,* strictly speaking, defines chemicals which are capable of promoting emulsification or stabilisation of emulsion or foams by interfacial action. In other food applications the term *emulsifier* may be misleading, since the actual function in a product may not be at all connected with emulsification, but with interaction with other food components such as starch and protein.

The world of surface activity and surface-active agents can appear complex and confusing to those not involved in it on a day-to-day basis. Most scientists and technical professionals and students in general courses have very little familiarity or training with the nature of surfaces, the properties of surface-active material and surfactants. They often have problems, therefore, recognising the need for such materials in their product development and processes or in identifying the surfactant properties that will satisfy their needs. In most cases, it is difficult for the non-specialist to correlate the available information in a way that facilitates practical application to a specific system or process.

The aim of this chapter is to provide a balance between the basic principle aspects of surfactant science and the less well defined and more empirical or practical world of its application to aid food technologists and research and development personnel to make educated guesses as to a preferred material or materials for a specific application. This may help avoid lengthy and expensive experimental work.

8.2 Classification, regulation and function

Emulsifiers or surfactants can be classified in different ways.

(A) Based on charge

Anionic Surfactants that carry a negative charge on the active portion of the molecule.

Cationic Surfactants carrying a positive charge on the active portion of the molecule.

Non-ionic Uncharged molecule having lipophilic and hydrophilic segments (such as a monoglyceride of a long-chain fatty acid).

Amphoteric Surfactant species that can be either cationic or anionic depending upon the pH of the solution.

Zwitterionic Both positive and negative charge may be present in the surface-active portion.

(B) Based on hydrophilic–lipophilic balance (HLB)

(C) Based on solubility.

(D) Based on functional groups: saturated/unsaturated, acids, alcohols, ethylene oxide, etc.

All food emulsifiers used in the United States and in Canada are regulated by the Food and Drug Administration (US FDA 21CFR, 182, 183) and by the Canadian Food and Drug Act respectively. In Japan the use is strictly regulated and it has to be publicly announced as food additives. In Europe, emulsifiers are regulated by the European Food Emulsifier Manufacturers Association (EFEMA) and given an E number. South Korea and Taiwan also have some type of approval list for each surfactant. Many other countries utilise the FAO/WHO Codex Alimentarius Commission which classifies emulsifiers as food additives. Lecithin, mono- and diglycerides and diacetyl tartaric acid esters of mono- and diglycerides have been given a GRAS status and therefore can be used in almost any food product and levels intended to achieve the intended technical or physical effects. The other emulsifiers are all regulated substances, meaning they have specific regulations that permit their use in specific products at specific use levels. Some of the regulations are more detailed than others.

In the US food additives are defined in one of the following categories.

GRAS (Generally recognised as safe) Additives which have been in use for many years. Their safety is generally based on extensive toxicological test data or based on use experience for an extended period of time. Their use may be limited.

GMP (Good Manufacturing Practice) Use not limited as to quantity or application.

Restricted Use limited to specified levels in specific applications.

Indirect Not added directly to food but may contact food via packaging or processing such as pan release aid in baking.

Incidental Emulsifiers used as a functional additive in an ingredient but not functional in secondary application.

Emulsifiers in foods can perform a variety of functions, the most important involves the reductions of surface tension in oil–water interfaces, interactions with starch and protein components and modification of the crystallisation of fats and oils. An emulsion is basically the dispersion of small droplets of one insoluble substance within another. Emulsions are usually the first type of mixture considered when talking about food emulsifiers. Selection of a proper type of emulsifier for a given food product is normally based on experience and experimental tests or by trial and error. In food applications, surfactants exert a variety of specific functions, other effects can also be obtained, many of these uses may overlap in any given application. It is essential that correct choice of emulsifier type and dosage be made to ensure optimum performance. A brief description of the major functions of surfactants in food is described below.

(a) *Emulsification*. All emulsifiers are surface-active agents which can promote emulsification of oil and water phases because they possess both hydrophilic and lipophilic groups within the same molecule. Emulsifiers with low hydrophilic to lipophilic ratio value (HLB value) stabilise water in oil (W/O) emulsions, whereas emulsifiers with high HLB values stabilise oil in water (O/W) emulsions. Each system requiring an emulsifier has an optimum HLB value which can be determined experimentally. This HLB value can then be used as an approximate indication of the most suitable emulsifier for that particular system. However, the chemical type of the emulsifier can often be of equal importance in achieving optimum emulsion stability. This function can be seen in margarine, coffee whiteners and creamy salad dressing.

(b) *Starch complexing (antistaling)*. Most emulsifiers containing a straight fatty acid chain in the molecule are capable of complexing with the soluble amylose component of starch. This property is of great importance in retarding staling in bread and cake and reducing stickiness in reconstituted starch based products such as instant mashed potato and pasta.

(c) *Protein interaction*. Emulsifiers possessing an ionic structure can interact with proteins in a number of food products to give improved structural properties. In bread they are capable of interacting with wheat gluten, giving rise to greater elasticity of the protein, therefore resulting in enhanced loaf volume.

(d) *Viscosity modifiers*. Certain emulsifiers when added to food products which contain sugar crystals dispersed in fat are capable of reducing the viscosity of the system by forming a coating on the sugar crystals. This property is extremely useful in modifying the flow properties of molten chocolate.

(e) *Foaming and aeration.* Emulsifiers containing saturated fatty acid chains are capable of stabilising aqueous foams and are therefore useful whipping agents in instant desserts, cake mixes, whipped toppings and other convenience foods. Emulsifiers containing unsaturated fatty acid chains are capable of depressing foam and are therefore useful as antifoaming agents in dairy products and egg processing.

(f) *Texture modification.* This is a complexing action on starch which reduces clumping and improves consistency and uniformity as in macaroni, dehydrated potatoes, breads and cakes.

(g) *Lubrication.* Emulsifiers such as saturated mono- and diglycerides tend to possess exceptionally good lubricating qualities in extruded starch products by lubricating the die thus giving greater process control. Addition of 0.5–1% of mono- and diglycerides in caramels will reduce their tendency to stick to cutting knives, wrappers and the consumers' teeth. Surfactants are also used to reduce stickiness or tackiness in products such as candies and chewing gum.

(h) *Crystal modification.* The modification of the polymorphic form, size and rate of growth in fat crystals such as margarine, shortening, chocolate and peanut butter can be influenced by certain emulsifiers in conjunction with optimised processing conditions. Optimisation of crystal size of such fats gives enhanced creaming properties and improved baking performance. This function can also apply to sugar and salt systems (sodium chloride).

(i) *Wetting.* Surfactants are generally effective wetting agents. However, the choice of surfactant is governed by the type of wetting to be accomplished such as wetting of waxy surface, capillary wetting or powder wetting. The emulsifier will function by the reduction of the interfacial tension between liquid and solid surfaces to cause a liquid to spread more quickly and evenly over surfaces as used in convenience foods such as spray-dried dessert mixes, coffee whiteners, drink mixes, instant breakfasts, rehydration of coconut or dry vegetables.

(j) *Solubilising.* Improving the ability of liquid in liquid dispersions to form clear solutions. Various colour and flavour systems require emulsifiers for solubilisation.

(k) *Demulsification – antifoaming.* Formulator is generally concerned with making a product stable but at certain applications, particularly during processing, breaking of an emulsion or a foam is desired. In most cases where demulsification is desired, an emulsifier of the opposite type or one that will throw the emulsifier system out of balance is used. The selection of surfactants will depend on the type of foam. In foods such as ice-cream, a carefully selected emulsifier produces controlled demulsification. It helps fat particles agglomerate to the size that best produce a dry product.

(l) *Palatability improvement.* This is the emulsification of a lipid system to enhance eating quality such as in chewing gum, icings and confectionery coatings.

(m) *Suspensions.* Suspensions are stable dispersions of finely divided insoluble material in a liquid medium. The size of the dispersed particles may vary from 0.1 μm to flocculates or aggregates as large as 100 μm. Surfactants are used in suspensions to aid the wetting of the insoluble particles. In turn, this helps ensure uniformity of the product. In products where separation occurs during storage, the surfactant helps redisperse the insoluble ingredients when the product is remixed. In suspensions, surfactants are generally used together with stabilisers or thickeners to produce the desired result. Chocolate drinks are among the most common suspensions in the food industry.

(n) *Dispersion.* Dispersions of solids, liquids and gases depend on the reduction of interfacial energy by a surface-active agent. The disperse systems can involve all three principal phases, as illustrated below. Many food applications such as ice-cream, coffee whiteners and flavoured beverages utilise this phenomenon.

Type	Internal phase	External phase
Emulsion	Liquid	Liquid
Foam	Gas	Liquid
Aerosol	Liquid or solid	Gas
Suspension	Solid	Liquid

8.3 Selection of surfactants and the HLB concept

The most important theme in the study of emulsions, solubilisation and other emulsifier related functions is the selection of a suitable surfactant that will satisfactorily perform the desired role at a given set of conditions. For food applications, the selection of the proper surfactant to provide the best performance from among the approximately 35 different chemical groups that have been approved as food additives and their combinations by the process of elimination is a formidable task. The stability of any emulsion is provided by a proper emulsifying agent which is adsorbed at the interface. The orientation of molecules with hydrophilic and lipophilic properties at the interface is, in terms of energy, more stable than in either the oil or the water phase. Surfactants tend to reduce surface tension between two immiscible liquids and the energy needed for emulsification of the two liquids is therefore reduced. The stability of an emulsion as shown in Table 8.2 is dependent on several factors:

1. Viscosity of the continuous phase.
2. Electric charge.
3. Adsorption of solid particles to the surface of the emulsified phase.
4. Formation of a mono- or multimolecular layer at the interface between the emulsifier phase and the continuous phase due to surfactant addition.

Table 8.2 Relation of emulsion stability and appearance with particle size

Particle size (μm)	Appearance	Stability
0.05 and smaller	Transparent	Extremely stable
0.05–0.1	Translucent	Excellent
0.1–1.0	Blue white	Good
1.0–10.0	Milky white	Tendency to cream
>10.0	Coarse	Quick breaking

Emulsion stability can be studied by various methods such as the determination of the rate of globule coalescence by particle size analysis or by phase inversion temperature (PIT) which can be determined by differential thermal analysis. Good correlation has been reported between PIT values and emulsion stability measured by particle size, also HLB number and HLB temperatures were found interrelated as long as the specification of the system is known.

Any attempt to simplify surfactant selection requires a method of classification. The most important property of surfactants is their polarity or hydrophilic–lipophilic ratio. The concept of classification of non-ionics based on the HLB value has contributed significantly to the systematic selection of surfactants.

In the HLB system of surfactant classification, surface-active agents are assigned values of 0–20. Oil-soluble materials are rated at the lower end of the scale while higher values are assigned to those materials exhibiting the greatest affinity for water. A surfactant with an equal affinity for water and oil is assigned a value of 10.

When a product is 100% hydrophilic, then it is arbitrarily assigned a value of 20, therefore the HLB value is basically an indication of the percentage weight of the hydrophilic portion of a non-ionic emulsifier molecule. The HLB numbers for most non-ionic emulsifiers can be calculated from either theoretical composition (molecular weight) or analytical data such as saponification value and acid number. The latter, where the data are available, provides a more accurate determination.

The calculation, based on molecular formula of the emulsifier, could cause some problem because industrially produced surfactants are never completely pure but are always composed of many homologous esters.

A rough approximation of HLB number may also be obtained by the apparent water solubility of the agent. Exceptions to these relations are known, but solubility serves as a quick means of approximating the HLB value of materials whose surfactant effects are unknown, as shown in Table 8.3. Another method of determining HLB number is based on functional groups of the surfactants such as polarity, hydration and electric charge.

Table 8.3 Correlation of HLB and water solubility or dispersibility

Behaviour of surfactant in water	HLB range
No dispersibility in water	1–4
Poor dispersion	3–6
Milky dispersion after vigorous agitation	6–8
Stable milky dispersion	8–10
Translucent to clear dispersion	10–13
Clear solution	13+

The HLB system can be used to assist surfactant selection in simple systems on the basis that every application has an optimum or required HLB. The optimum emulsion stability of W/O and O/W emulsions was found at the HLB numbers of 3.5 and 12 respectively. Outside these HLB values, the rate of coalescence was sharply increased. It was also demonstrated that O/W emulsions prepared with blends of emulsifiers are more stable than emulsions prepared with a single emulsifier providing the HLB numbers are maintained.

The principle is illustrated where performance is plotted against HLB for three different classes of emulsifiers. If the required HLB can be established experimentally by employing one type of emulsifier, such as a blend of sorbitan monostearate and polysorbate 60, then further trials can be limited to surfactants with HLB numbers near the established optimum. Chemical type can be a major factor in performance, so that the effect of a change should be thoroughly studied in arriving at the preferred emulsifier. In this case, the emulsifiers made from the unsaturated oleic acid show improved performance. It should be noted that this optimum performance is found in the same HLB range.

There are some 'rules-of-thumb' that can also be used as an aid in emulsifier selection. There is a relationship between the solubility of an emulsifier and its behaviour (Table 8.3). Thus, a water-soluble emulsifier or blend is required to produce an O/W emulsion, to solubilise oils, or to obtain detergent action. On the other hand, in order to produce W/O emulsions, an oil-soluble emulsifier with a low HLB is indicated, as illustrated in Table 8.4

Table 8.4 HLB as an indication of surfactant function

HLB	Function
4–6	W/O emulsifiers
7–9	Wetting agents
8–18	O/W emulsifiers
13–15	Detergents
15–18	Solubilisers

Sometimes the HLB concept may provide only a partial answer to selecting a surface-active agent. In a formula composed of oil, water and emulsifier, correlations are possible in most instances. However, the addition of flour, starch, sugar, milk, salt, eggs and similar ingredients, most of which contain natural emulsifier, causes some complications. Thus, the formulation of new food products with complex systems is often faced with a combination of HLB studies coupled with the empirical approach. It is common to evaluate surfactants which have shown efficiency in the same or similar applications in test formulas in the hope of finding a quick, satisfactory solution. The inherent fallacy of this system has been demonstrated many times. In complex surfactant projects where there are many variables and the experimental errors in the test method are large – as in the test baking of bread or cakes – the problem of selecting the best combination of emulsifiers from among so great a number is extremely time consuming. As a result, more sophisticated methods have been investigated and statistical methods were found to reduce the laboratory effort required to run empirical evaluations.

The statistical techniques using a factorial design, where all variables are changed simultaneously, could help in selecting the proper emulsifier for a particular application. Such information may help reduce significantly the amount of laboratory experimentation. The availability of computer programs has also enabled the use of the response surface methodology with experimental designs to achieve and explore all possible combinations with fewer experimentations. These methods are becoming widespread as the need for complex emulsifier systems becomes apparent in the future. The required HLB numbers for selected oils used in food applications are listed in Table 8.5.

Table 8.5 Required HLB numbers for emulsification of common oils used in food application

Compound	HLB numbers	Compound	HLB numbers
Lauric acid	16	Mineral oil, aromatic	12
Linoleic acid	16	Mineral oil, paraffinic	10
Oleic acid	17	Mineral spirits	14
Ricinoleic acid	16	Palm oil	7–10
Beeswax	9	Paraffin wax	10
Castor oil	14	Rapeseed oil	9
Cocoa butter	6	Safflower oil	7
Corn oil	8–10	Soybean oil	6
Cottonseed oil	5–6	Tallow	6
Lard	5	Hydrogenated peanut oil	6–7
Menhaden oil	12		

8.4 Surfactants used in food

The surfactants that are permitted for use as food additives are limited in number because countries around the world are passing food laws that will allow the use of approved and demonstrably safe emulsifiers only. These measures are important primarily for protecting the health of the consumer and also to simplify international food trade. The most common emulsifiers with their HLB value, acceptable daily intake (ADI) and regulation status are shown in Table 8.6. The Food Chemical Codex specification for food emulsifiers is presented in Table 8.7.

8.4.1 *Monoglycerides*

Monoglycerides are the most commonly used food emulsifiers with GRAS status. They are lipophilic, non-ionic and manufactured by esterification or alcoholysis of edible fats and oils or fatty acids with glycerol in the presence of a catalyst.

Distilled monoglycerides of 90% monoester and above are obtained by molecular distillation. Monoglycerides are usually present in a mixture of α- and β-forms. The α-crystalline form is the most active in terms of functionality.

The α-mono content can vary from 40 to 90% with an iodine value of 1–100 and a melting point of approximately 40–70°C. They are available in powder, flakes, plastic and liquid form.

Monoglycerides function as emulsifiers and stabilisers in several foods such as ice-cream, baked goods, shortening, margarine, toppings, whiteners, pasta, confectionery, processed potatoes, multivitamin preparation and snack food. Distilled monoglycerides are also used as an excellent raw material for the manufacture of lactic acid, tartaric, acetic and citric acid

Table 8.6 HLB number ADI, regulation status, EEC number and US FDA Regulation Code for Food Emulsifiers

Emulsifiers	HLB	EEC No.	ADI[1]	Regulation status	US FDA 21 CFR
Lecithin	3–4	E322	Not limited	GRAS	184.1400
Modified lecithin	10–12	—	No ADI	GRAS	172.814
Mono- and diglycerides	3–4	E471	Not limited	GRAS	182.4505
Ethoxylated monoglyceride	10–12	—	—	Regulated	172.834
Sorbitan monostearate	3–6	491	0–25	Regulated	172.842
Polysorbate 60	14.4	435	0–25	Regulated	172.836
Polysorbate 65	10–11	436	0–25	Regulated	172.838
Polysorbate 80	15.4	433	0–25	Regulated	172.840
Succinylated monoglycerides	5–7	—	—	Regulated	172.830
Sucrose esters	3–16	E473	0–2.5	GMP	172.859
Lactic acid esters of mono- and diglycerides	3–4	E472b	Not limited	GMP	172.852
Polyglycerol esters of fatty acid	5–13	E475	0–25	GMP	172.854
Propylene glycol esters of fatty acid	2–3	E477	0–25	GMP	172.856
Sodium stearoyl lactylate	10–12	E481	0–20	Regulated	172.846
Calcium stearoyl lactylate	5–6	E482	0–20	Regulated	172.844
Diacetyl tartaric acid esters of mono- and diglycerides	8-10	E472e	0–50	GRAS	182.4101
Acetic acid esters of mono- and diglycerides	2–3	E472a	Not limited	GMP	172.828
Citric acid esters of mono- and diglycerides	10–12	E472c	Not limited	—	172.832
Salts of fatty acid (Na, K)	16–18	E470	Not limited	—	172.863

[1]ADI – acceptable daily intake (mg kg^{-1} body weight)

esters. The approximate composition of a typical distilled high and low monoglyceride content is shown in Table 8.8.

Many types of monoglycerides are available. They are made from different sources of fats or oils and therefore their iodine value, melting point and form is different, as shown in Table 8.9, which also gives common mixtures of monoglycerides, polysorbate and sorbitan esters and their characteristics.

Table 8.7 Surfactants listed in the Food Chemical Codex (FCC) and their chemical specification

Product	Functional use in food	Acid value	Hydroxyl value	Residue on ignition, (%)	Saponification value	Water (%)	Additional
Acetylated monoglycerides	Emulsifier, coating agent, texture modifying agent, solvent, lubricant	6 max.			Vendor spec.		Reichert-Meissl value 75–150[1], vendor spec. for iodine value and free glycerol
Calcium stearoyl lactylate	Dough conditioner, stabiliser, whipping agent	50–86					Calcium content 4.2–5.2%; ester value 125–164; total lactic acid 32–38%[1]
Diacetyl tartaric acid esters of mono- and diglycerides	Emulsifier	62–76		0.01 max.	380–425		Tartaric acid 17–20%; acetic acid 14–17%; fatty acid 56.0% min. (after saponification); glycerol 12 min. (after saponification)[1]
Ethoxylated mono- and diglycerides	Dough conditioner, emulsifier	2 max.	65–80		65–75	1 max.	1,4-dioxane – passes test; oxyethylene content (apparent) 60.5–65.0%; stearic, palmitic and myristic acid 31–33%[1]
Lactated mono- and diglycerides	Emulsifier, stabiliser	Vendor spec.	Vendor spec.			Vendor spec.	1-Monoglyceride content; total lactic acid, free glycerol-vendor spec.[1]
Lecithin	Antioxidant, emulsifier	36 max.				1.5 max.	Acetone-insoluble matter 50% min; heavy metals (as Pb) 0.004 max.; hexane-insoluble matter 0.3% max.; lead 10 ppm max.; peroxide value 100 max.
Mono- and diglycerides	Emulsifier, stabiliser	6 max.	Vendor spec.	0.5 max.	Vendor spec.		Free glycerine 7% max.; α-monoglyceride, iodine value; total monoglycerides, hydroxyl value – vendor spec.[1]

Name	Function						Specification
Monoglyceride citrate	Synergist solubiliser	70–100		0.3 max.	260–265	0.2 max.	Total citric acid 14–17%[1]
Polyglycerol esters of fatty acid	Emulsifier, aeration, defoaming, clouding agent, lubricant	Vendor spec.	Vendor spec.	Vendor spec.	Vendor spec.		Iodine value, sodium salt of fatty acid – vendor spec.[1]
Polysorbate 20	Emulsifier, stabiliser, aeration, defoaming, crystal control, wetting	2 max.	96–108	0.25	40–50	3 max.	Assay for oxyethylene content 70–74%; 1,4-dioxane – passes test; lauric acid 15–17 g per 100 g[1]
Polysorbate 60	Emulsifier, stabiliser, aeration, defoaming, crystal control, wetting	2 max.	81–96	0.25	45–55	3 max.	Assay for oxyethylene content 65–69.5%[1]; 1,4-dioxane – passes test; stearic and palmitic acid 24–26 g per 100 g
Polysorbate 65	Emulsifier, stabiliser, aeration, defoaming, crystal control, wetting	2 max.	44–60	0.25	88–98	3 max.	Assay for oxyethylene content 46–50%; 1,4-dioxane – passes test; stearic and palmitic acid 42–44 g per 100 g[1]
Polysorbate 80	Emulsifier, stabiliser, aeration, defoaming, crystal control, wetting	2 max.	65–80	0.25	45–55	3 max.	Assay for oxyethylene content 65–69.5%; 1,4-dioxane – passes test; oleic acid 22–24 g per 100 g[1]
Propylene glycol mono diesters	Emulsifier, stabiliser	4 max.		0.5			Free propylene glycol 1.5% max.; soap (as potassium stearate) 7% max.; total monoester content – vendor spec.[1]
Sodium stearoyl lactylate	Emulsifier, dough conditioner, stabiliser, whipping agent	60–80					Ester value 150–190; Sodium content 3.5–5%; total lactic acid 31–34%[1]

Table 8.7 *Continued*

Product	Functional use in food	Acid value	Hydroxyl value	Residue on ignition, (%)	Saponification value	Water (%)	Additional
Sodium stearoyl fumarate	Dough conditioner				142.2–146	5.0 max.	Assay 99–101.5 $C_{22}H_{39}NaO_4$; lead 10 ppm; heavy metals 0.002% max.; sodium stearoyl maleate 0.25% max.; stearyl alcohol 0.5 max.
Sorbitan monostearate	Defoaming agent, emulsifier, stabiliser	5–10	235–260		147–157	1.5 max.	Assay 29.5–33.5 g per 100 g of polyols 71–75 g per 100 g fatty acid
Stearoyl monoglyceridyl citrate	Emulsion stabiliser	40–52		0.1 max.	125–255	0.25 max.	Total citric acid 15–18%[1]
Succinylated monoglycerides	Emulsifier, dough conditioner	70–120	138–152				Bound succinic acid 14.8 min.; free succinic acid 3% max.; iodine value 3 max.; total succinic acid 14.8–25.6%[1]
Sucrose esters	Emulsification, starch complexing, aeration	5 max.				4 max.	Free sucrose 2% max.

[1] Heavy metals value of 10 ppm max.; arsenic value of 3 ppm max.

Table 8.8 Typical stearic acid based monoglycerides composition

Composition	Distilled	High mono	Low mono
Total monoglycerides	95.0	65	50
α-Monoglycerides	92.0	60	40–45
Diglycerides	3.5	34	45–50
Triglycerides	0.1	3.5	5–10
Free fatty acids	0.8	1	1–2
Free glycerol	0.7	1.5	2–5
Approximate fatty acid distribution*			
Myristic acid (14:0)	3	3	1
Palmitic acid (16:0)	33	30	10
Stearic acid (18:0)	62	65	85–90
Oleic (18:1)	<2	<2	1
Arachidic acid (20:0)	2	2	1

*The fatty acid composition may vary depending on source of raw material used.

Table 8.9 Characteristics of different types of mono- and diglycerides and their blends with polysorbate and sorbitan ester

Surfactant	Form	Melting point vis. at 25°C	HLB rating	Iodine value
Mono- and diglycerides from the glycerolysis of edible fats or oils (52–56% α-mono, 61–66% total mono)	Bead form	57–61	3.2	<2
Mono- and diglycerides of fat forming fatty acids (47–50% α-mono, 54–59% total mono)	Liquid	App. 150 cP	2.8	74–78
Mono- and diglycerides from the glycerolysis of edible fats or oils (40–44% α-mono, 48–52% total mono)	Votated plastic solid	46–50	2.8	56–64
Mono- and diglycerides from the glycerolysis of edible fats or oils (40–44% α-mono, 48–52% total mono)	Votated plastic solid	50–51	2.8	52–60
Mono- and diglycerides from the glycerolysis of edible fats or oils (40–44% α-mono, 48–52% total mono)	Bead or flake	57–61	2.8	<3
Mono- and diglycerides from the glycerolysis of edible fats or oils (40–44% α-mono, 48–52% total mono)	Votated plastic solid	33–35	2.8	<8
Mono- and diglycerides from the glycerolysis of edible fats or oils (54–58% α-mono, 65–69% total mono)	Votated plastic solid	51–53	3.5	66–72
Mono- and diglycerides from the glycerolysis of edible fats or oils (52–56% α-mono, 61–66% total mono)	Bead form	60–62	3.5	54–61
Mono- and diglycerides from the glycerolysis of edible fats or oils (54–58% α-mono, 65–69% total mono)	Votated plastic solid	54–57	3.5	<2

Table 8.9 *Continued*

Surfactant	Form	Melting point vis. at 25°C.	HLB rating	Iodine value
Mono- and diglycerides from the glycerolysis of edible fats or oils (54–58% α-mono, 65–69% total mono)	Votated plastic solid	53–56	3.5	46–50
Stearic acid monoglyceride, distilled 95% min. α-mono	Powder	63–68	3.8	2 max.
Oleic acid monoglyceride, distilled min. 93% α-mono	Solid	40–45	3.8	70–80
Fully hydrogenated vegetable oil, distilled mono- and diglyceride 95% min. α-mono	Powder and beads	63–68	3.8	2 max.
Sorbitan monostearate	Bead	52.8	4.7	71–75
Polyoxyethylene (20) sorbitan monostearate	Yellow oily liquid (may gel on standing)	600 cP	14.9	—
Polyoxyethylene (20) sorbitan tristearate	Tan waxy solid	App. 33.3	10.5	—
Polyoxyethylene (20) sorbitan monooleate (Polysorbate 80)	Yellow oily liquid	400 cP	15.0	—
Mono- and diglycerides from the glycerolysis of edible fats or oils (80%) and polysorbate 80 (20%)	Solid	54–57	5.2	App. 6
Mono- and diglycerides from the glycerolysis of edible fats or oils (80%) and polyoxyethylene (20) sorbitan tristearate (20%)	Bead form	54–57	4.3	—
Mono- and diglycerides from the glycerolysis of edible fats or oils (40%) and polyoxyethylene (20) sorbitan tristearate (60%)	Bead	54–57	5.9	—

8.4.2 *Ethoxylated mono- and diglycerides*

Hydrophilic and non-ionic emulsifiers manufactured by the reaction of ethylene oxide with low mono- and diglycerides under pressure. They function as emulsifiers and aerating agents and are used in baked products, toppings, whiteners, non-standarised frozen desserts and icings.

8.4.3 *Sorbitan esters*

Lipophilic, non-ionic emulsifiers made by the reaction of polyhydric alcohol (sorbitol) and stearic acid. The most common products used in food applications are sorbitan monostearate and sorbitan tristearate. They function as emulsifiers, aerating agents and lubricants in cakes, toppings, whiteners, confectionery coatings, active dry yeast and as crystal modifiers in margarine.

8.4.4 Polysorbates

This group of products consists of polysorbate 20, 60, 65 and 80. They are non-ionic, hydrophilic surfactants and are obtained by high-pressure reaction of sorbitan esters with ethylene oxide. They provide several functions. They are used in baked products, toppings, icings, ice-cream, whiteners, dressings, flavours, confectionery, coating and shortening.

8.4.5 Sucrose esters

These are common surfactants used in Japan and the Far East. They recently were approved by the US FDA for use in food products. They are non-ionic emulsifiers with variable HLB numbers. They are obtained by reacting sucrose with fatty acid methyl esters. The final reacted product is further purified. These types of emulsifiers are used mainly in O/W emulsions and as a basis for food, detergents, crystallisation adjuster, anti-spattering and foaming agents. They are used in such foods as baked goods, noodles, dairy analogues, frozen desserts and fresh fruit coating.

8.4.6 Citric acid esters of mono- and diglycerides

Obtained by the reaction of citric acid with distilled or regular mono- and diglyceride. They are used mainly as antioxidant improvers, coalescents in meat emulsions and emulsifiers for shortening. Monoglyceride citrates are also more powerful sequestering agents for trace metals than citric acid.

8.4.7 Polyglycerol esters of fatty acid

Emulsifiers produced by polymerising glycerol and then esterifying the glycerol polymer with selected fatty acids (such as palmitic, stearic and oleic). These esters are non-ionic and possess both hydrophilic and lipophilic properties. They are used mostly in icings, toppings and cakes. Polyglycerol esters have been claimed to replace polysorbate in certain food applications.

8.4.8 Propylene glycol monostearate

Lipophilic and non-ionic emulsifiers manufactured by the reaction of propylene glycol with stearic acid (PGMS). The other type, PGME, are obtained by the reaction of fat with glycerol and propylene glycol. The difference between the two types is illustrated below.

	PGMS	*PGME*
Monoester content (%)	45–90	35–65
Iodine value	5 max.	1–60
Form	Solid, flakes	Solid, flakes, plastic
α-Monoglyceride (%)	—	10–20

They are used mainly in cakes and toppings as emulsifiers and aerating agents.

8.4.9 *Lactic acid esters*

Lactic acid esters are manufactured by reacting stearic acid (50–90%, C:18) with food grade lactic acid. The salt is obtained by neutralisation with sodium or calcium hydroxide. Lactic acid esters are commercially very important food emulsifiers. They can be divided into two groups, anionic and non-ionic esters. Non-ionic esters are widely used in bakery and toppings but they do not form mesomorphic phases with water. The most important of the anionic group are calcium stearoyl lactylate (CSL), sodium stearoyl lactylate (SSL) and stearoyl lactoyl lactic acid (SLA). Their main use is in baked products, whipping agents, dehydrated potatoes, puddings and whiteners. They are permitted in the US to be used at 0.5% of flour weight in white pan bread. In Canada, their maximum use in white bread is 0.375% of flour weight.

8.4.10 *Lecithins*

Lecithins continue to play an important role in many food formulations as a natural emulsifier and dispersing agent. Worldwide production is estimated to be over 100 000 tonnes annually. Most of the lecithins produced commercially are obtained from soybeans. Other oilseeds such as cottonseed, corn, rapeseed and sunflower represent new commercial sources of this valuable ingredient. Several types of lecithins and modified lecithins are available to the food formulators and can exist as liquid, plastic or free flowing solid.

(a) *Oil Free* Granules obtained by acetone extraction, followed by vacuum drying. It contains 2–3% soya oil to preserve colour and odour.
(b) *Fractionated* Ratio of phosphatides changed to alter emulsification properties. This can be achieved chemically or enzymically.
(c) *Hydroxylated lecithin* Unsaturated sites of lecithin fatty acid treated with hydrogen peroxide and lactic acid to increase hydrophilic characteristics.

Lechithins are used extensively in many types of foods such as yeast-raised bakery products, cakes, chocolate, pan release aids and others to perform a variety of functions. A comparison between the general characteristics of natural and hydroxylated lecithins is shown in Table 8.10.

Table 8.10 Characteristics of lecithins

Characteristics	Natural	Hydroxylated
Acetone insoluble (%)	54–72	60 min.
Moisture (%)	0.75 max.	1.5 max.
Iodine value	95–98	90 max.
Approximate HLB	4	12

8.4.11 Phosphatides

Natural emulsifiers obtained from vegetable oil and egg yolks. They possess lipophilic characteristics and are amphoteric in nature (anionic and cationic). They are considered GRAS. This type of emulsifiers are a mixture of lecithin, cephalin and inositol and are obtained by drying a water emulsion of an oil. They supply emulsification, lubrication and wetting properties in margarine, chocolate, candy, ice-cream, baked products and fats and oils. Also, they control viscosity of molten chocolate.

8.4.12 Diacetyl tartaric acid esters of mono- and diglycerides (DATAE)

Hydrophilic, non-ionic emulsifiers which have gained an important role in the food industry. DATA ester is obtained by the reaction at 100–130°C between a monoglyceride (preferably distilled) and diacetyl tartaric acid anhydride (made from 1 mole tartaric acid and 3 moles acetic anhydride). The physical properties of DATAE depend primarily on the type of fatty acids and the quantity of esterified tartaric acid compared with esterified acid. DATAE are hygroscopic and have a range of melting points between 20 and 50°C. They function as emulsifiers and starch and protein complexing agents. They are excellent O/W emulsifying agents and are used in feeding emulsions, antispattering agents, icings, foam-stabilising agents and dough-conditioning agents.

8.4.13 Succinylated monoglycerides

Lipophilic, non-ionic obtained by the reaction of succinic anhydride and monoglycerides. They contain 60–75% succinylated mono- and 12–20% free monoglyceride. They function as emulsifiers and starch and protein complexing. They are used in yeast-raised baked goods and in shortening.

8.4.14 *Lactylated monoglycerides*

Lipophilic, non-ionic emulsifiers and aerating agents obtained by the reaction of lactic acid with monoglycerides and are used in cakes, toppings and whiteners, they contain 7–12% α-monoglycerides.

8.5 Major applications of emulsifiers in food

8.5.1 *Ice-cream*

Ice-cream is considered a W/O emulsion and a foam. The most common emulsifiers used in ice-cream are lecithins, mono- and diglycerides, DATAE ester and Polysorbate 80. Recently, sucrose ester and polyglycerol ester have been reported to be functional in ice-cream. Stabilisers from different types of gum are used with emulsifiers to obtain stability and quality. The level of emulsifier used in ice-cream can vary from 0.1 to 0.35% depending on the fat level in the formula. Emulsifiers perform the following functions.

1. Promote the destabilising of the mix during freezing.
2. Act as nuclei for the crystallisation of high melting fats.
3. At certain temperatures, can form various mesophases at the interface.
4. Improve whipping quality of the mix (overrun).
5. More efficient air incorporation.
6. Production of a drier ice-cream.
7. Smoother body and texture.
8. Superior scooping qualities at the freezer.
9. Better control over various manufacturing processes (extruding and packaging).
10. Increased melt time.
11. Lower surface tension.
12. Reduction in whipping time.
13. Produce somewhat smaller ice crystals more evenly distributed and smaller air cells which result in a smoother ice cream.

Many factors could affect the action of an emulsifier in ice-cream such as ingredients used, procedure of processing, freezing, hardening, homogenisation and the level and type of emulsifier and stabilisers used.

8.5.2 *Surfactants in the baking industry*

Baked products require several types of surfactants to provide a variety of functions. They can be categorised in the following four distinct groups.

1. *Filling and icings.* Many bakery products rely on the use of fillings and icings, the structural properties of which can be suitably modified by the use of an appropriate emulsifier that provides the following functions:

(a) Controls syneresis (weeping)
(b) Increases volume
(c) Improves moisture retention
(d) Improves stiffness (body) of icing
(e) Improves appearance, texture and mouthfeel
(f) Reduces mixing time
(g) Aids aeration and improves yield
(h) Gives longer shelf-life

2. *Prepared mixes.* A range of cold water dispersible emulsifiers is available for use in cake mixes where rapid reconstitution and aeration are necessary. Various carrier-dispersed emulsifiers are also available for incorporation into prepared mixes where cold-water dispersibility is not required.

Other food items available in this category are mixes for muffins, doughnuts, biscuits and bread. These may contain one or a mixture of emulsifiers. Mono- and diglycerides, lecithin, PGME, PGMS, SSL, sorbitan ester, polysorbate and DATA esters are used.

3. *Cake improvers.* Emulsifiers can function in various ways to satisfy the numerous properties required in a wide range of cake formulations. The type and usage level of emulsifiers vary considerably depending on the type of cake produced. Emulsifiers generally provide the following function in cakes:

(a) Increase cake specific volume.
(b) Produce a more uniform texture and better crumb structure and create crumb softness.
(c) Reduce egg/shortening.
(d) Increase shelf-life by retarding staling.
(e) Impart tenderness and more rapid flavour release
(f) Reduce mixing time.
(g) Improve performance of dried egg for use in cake mixes.

4. *Surfactant in bread.* Bread staling is a term which indicates decreasing consumer acceptance of bakery products by changes in the crumb other than those resulting from action of spoilage organisms. The theories that can explain crumb staling are summarised as follows:

(a) Loss of moisture.
(b) Decrease in soluble starches and change in starch from the amorphous, partially gelatinised form to a less hydrated crystalline state with simultaneous liberation of moisture.

(c) Gradual and spontaneous aggregation of the amylopectin or branched fraction of starch giving rise to crystalline structure.
(d) Migration of moisture from gluten to starch phase or from starch to gluten.

Several changes in bread as a result of staling can be seen, such as:

(a) Increase in crumb firmness.
(b) Increase in crumbliness and harshness of texture.
(c) Deterioration in flavour and aroma.
(d) Loss of crust crispness, becoming soft and leathery.
(e) Increase in crystallinity of the crumb.
(f) Decrease in soluble starch.
(g) Decrease in susceptibility to β-amylase.

Dough conditioners, crumb softeners and emulsifiers are used in the production of yeast-raised baked goods to aid in production and/or improve certain quality factors. The major benefits of surfactant in the dough stage and the final products are summarised below.

Dough benefits

1. Gluten complexing to improve machinability (extensibility).
2. Improved gas retention.
3. Improved mixing tolerance.
4. Increased water absorption.
5. Improved handling stability and rate of proof.
6. Improved hydration rate of flour and other ingredients.

Product benefits

1. Improved loaf volume.
2. Bind with starch to retard the rate of crumb firming or staling (shelf-life)
3. Better texture and finer grain.
4. Improved symmetry.
5. Stronger side walls.
6. Improved slicing characteristics.
7. Shortening-sparing effect.

The factors that may be considered when choosing a surfactant are the type of bakery product produced, the desired effect and the best physical form of incorporation. Surfactants for the baking industry are available in votated, plastic bead, powder and liquid form. A wide range of surfactants of different chemical types and their combination can be utilised in all types of breads; and their usage level varies depending on function desired, regulation and type of product produced. A summary of the surfactants used in bread and their usage level is shown in Table 8.11. The antistaling (firming)

Table 8.11 Surfactants used in bread and their usage level

	Function		% Usage level in bread	% Usage level in non-standardised product
Dough conditioners/Crumb softeners	Conditioning	Softening		
Sodium stearoyl lactylate[2]	Excellent	Very good	0.5 max.[1]	GMP[5]
Calcium stearoyl lactylate	Excellent	Good	0.5 max.[1]	0.5 max.
DATA esters[3]	Excellent	Fair	GRAS	GMP
Ethoxylated monoglycerides	Very good	Poor	0.5 max.[1]	0.5 max.
Sucrose esters	Good	Good	0.5–1	GMP
Crumb softener/Dough conditioner				
Polysorbate 60[4]	Fair	Very good	0.5 max.[1]	0.5 max.
Succinylated monoglycerides	Good	Good	0.5 max.[1]	0.5 max.
Crumb softeners				
Distilled monoglycerides	Poor	Excellent	GRAS	GMP
Mono- and diglycerides	Poor	Excellent	GRAS	GMP
Soft mono- and diglycerides	Poor	Very Good	GRAS	GMP

[1] The total permitted alone or in combination cannot exceed 0.5% based on flour weight.
[2] In Canada the maximum is 0.375 based on flour weight.
[3] Used in Canada at 0.6% based on flour weight.
[4] Not permitted in Canada for bread.
[5] GMP = Good Manufacturing Practice.

Table 8.12 Comparison between the effect of different surfactants on specific volume and compressibility of white bread using vegetable shortening

Surfactants	Level[1] (%)	Specific volume (ml g^{-1})	Compressibility, g force[2] storage time (hours)			Softeness Index (hours)	
			24	96	168	96	168
Control	0.0	6.05	130	227	275	1.0	1.0
Distilled monoglycerides	0.375	6.04	111	207	248	0.91	0.90
Sodium stearoyl lactylate	0.375	6.85	66	124	171	0.54	0.62
Blend of SSL and monoglycerides	0.375	6.18	120	186	220	0.82	0.80
Calcium stearoyl lactylate	0.375	6.70	66	121	162	0.53	0.59
Diacetyl tartaric acid esters of monoglycerides	0.6	6.53	69	108	162	0.48	0.59
Lecithin	1	6.51	81	139	191	0.61	0.69
Blend of distilled monoglycerides and SSL	0.5	7.25	53	100	145	0.44	0.52
Mono- and diglycerides with 54–58% α-mono (plastic)	0.5	6.70	105	150	180	0.66	0.65

[1]Level of surfactant used is based on 100% of flour weight.
[2]Data obtained using the official AACC method using Baker Compressimeter.

characteristics provided by different surfactants on white pan bread are shown in Table 8.12. These unpublished results, based on work performed in the author's laboratory, may vary depending on baking conditions and formula used. The different surfactants utilised clearly demonstrated their advantages in reducing firmness, increasing shelf-life and improving volume in bread.

8.5.3 Emulsifiers in margarine and shortenings

Table margarines and low-calorie spreads are emulsions of water in liquid and solid fat blends. Small quantities of monoglycerides are added during the manufacture of the margarine to obtain fine dispersion of the water. Where large volumes of water are involved, as in the case of low-calorie spreads, monoglycerides containing unsaturated fatty acids are recommended.

Margarines and shortenings manufactured for use in bakery applications differ from domestic margarines. They require emulsifiers to give good creaming properties which are essential for optimum baking performance. Emulsifiers can also be supplied to suit specialised application requirements such as antispattering agents in frying margarines.

Several types of shortening are available, the most common of which are:

(a) Baker's shortening (multipurpose)
(b) Bread and sweet dough
(c) Fluid shortening
(d) Consumer shortening
(e) Cake mix shortening
(f) Icing and filling shortening

The different types of shortening utilise different types of emulsifiers. The usage levels can vary from 2 to 14%. A single emulsifier or a blend of several types may be used. The major type of emulsifiers used include different types of mono- and diglycerides varying in α-mono content, polysorbate, sorbitan esters, propylene glycol monostearate and polyglycerol esters.

8.5.4 Peanut butter

Peanut butter is made basically from roasted peanuts which contain approximately 50% oil. Salt, dextrose, emulsifiers and hydrogenated vegetable oil are added. The most common emulsifiers used are mono- and diglycerides of 52–90% monoesters. For consumer peanut butter the level can vary from 0.5 to 1.5%. The peanut butter used in confectionery has a high level of surfactant, up to 2.25%. In processing, emulsifiers can be mixed with dextrose and salt and added to dry peanuts or they may be melted and added to the mill. Emulsifiers in peanut butter can provide several benefits.

(a) Crystallise part of the free oil during processing and therefore improve stability during storage
(b) Improve gloss appearance
(c) Improve spreadability over a wide range of temperatures
(d) Provide production versatility
(e) Improve palatability.

Sesame spread, used widely in the Middle East, and other spreads such as hazelnuts or almonds can utilise emulsifiers to obtain similar benefits.

8.5.5 Emulsifiers in imitation dairy products

Many synthetic dairy products are essentially oil-in-water emulsions which depend upon emulsifiers to achieve satisfactory end-use properties.

In synthetic creams, monoglycerides and polyglycerol esters are extremely useful in attaining desired overrun, texture and palatability. Various blends of emulsifiers can be used to make specialty products such as coffee whiteners, fluid milks and calf milk replacers. Emulsifiers are useful wetting agents for incorporation in spray-dried milk-based products which are to be reconstituted with cold water. Specific emulsifiers such as propylene glycol esters and lactic esters of monoglycerides are extremely useful aerating agents in high fat content instant desserts. Coffee whiteners are the best example of imitation dairy products. They are based on vegetable or animal fat, and their usage level is increasing at a rapid rate because they offer lower cost and better shelf-life than conventional dairy products. Coffee whiteners are available in three physical forms: liquid, frozen and spray dried. Emulsifiers at usage levels of up to 0.5% play an important role in formulating such products. Other ingredients also must be balanced and be of good quality.

The most important factors to be considered to make good coffee whiteners are:

(a) *Stability.* A good uniform physical form with no oiling-off or feathering when added to coffee or tea.
(b) *Viscosity* – liquid type.
(c) *Whitening ability*, which depends on the amount of solids present and the fineness of the dispersed phase.
(d) *Flavour.* A bland flavour and odour-free product is highly desirable.
(e) *Flow properties.* A spray-dried product must exhibit good flow and should not cake or clump.
(f) *Easy dispersal.* A spray-dried product must disperse easily and be non-hygroscopic.

Mono- and diglycerides with an α-monoglyceride of 52–56%, SSL, polysorbate and sorbitan esters are among the best surfactants used in formulating different types of whiteners.

8.5.6 *Emulsifiers in starch products*

Starch-based products such as instant mashed potato and pasta benefit from the use of starch complexing emulsifiers which enhance their cooking stability. Extrusion of snack products and breakfast cereals is also facilitated by the use of starch-complexing emulsifiers.

8.5.7 *Emulsifiers in sugar confectionery*

Emulsifiers can modify the properties of sugar confectionery such as toffees and chewing gum. The presence of monoglycerides reduce stickiness in these products, so facilitating their processing and improving their palatability.

Many other applications for surfactants may be found in the literature. The data presented in Table 8.13 are a summary of the suggested use and benefit of surfactants in food. Emulsifiers can be applied singly or in combination to produce the best possible function. The levels suggested are only guides, and can be altered depending on the specific end product and the desired function. These emulsifier suggestions are also general, as the food regulations of some countries may not allow specific emulsifiers in some food items or they may not allow the usage levels suggested. Consulting regulations for specifics is always recommended.

8.6 Glossary

The following definitions of terms and phrases are very common in the science and applications of surfactants. In certain cases, two distinct groups, namely the academicians and industrial scientists, may have a slight disagreement on the meaning of the terms therefore these definitions may vary slightly from those found in other references. For the product development personnel and technologists, these terms are useful tools to assist in the understanding of the complex world of surfactants.

Biodegradability   A measure of the ability of a surfactant to be degraded to simpler molecular fragments by the action of biological processes.

Cloud point   For non-ionic surfactants – the temperature (or temperature range) at which the surfactant begins to lose sufficient water solubility to perform some or all of its normal functions as a surfactant.

Coalescence   The irreversible union of two or more drops or particles to produce a larger unit or lower interfacial area.

Diglyceride   A chemical combination of fatty acids and glycerine in the

Table 8.13 Emulsifiers suggested for selected processed foods – their function and usage levels

Emulsifier: (1) Mono- and diglycerides; (2) sorbitan esters; (3) sucrose esters; (4) PGME/PGMS; (5) lecithins; (6) SSL or CSL; (7) acetic, citric, lactic and tartaric acid monoglyceride; (8) DATA ester; (9) polyglycerol ester; (10) polysorbate.

Food	Function	Percent suggested usage level (based on total weight)[1]	(1)	(2)	(3)	(4)	(5)	(6)	(7)	(8)	(9)	(10)
Bread, rolls, and sweet goods	Retards crumb firming, dough conditioning reduce mixing time, increase absorption	0.2–0.5	X		X			X				X
Beverage mix (non-alc.)	Foaming agent	1.0–3.0									X	
Biscuits	Quality improver, anticrystalling	0.1–0.5	X					X				
Cakes	Improve volume, texture, batter aeration	1.5–2.5	X	X					X			X
Cake mixes	Whipping agent, quality improver	5–10[2]	X	X	X	X	X	X		X	X	X
Caramels	Improves chewing quality, reduces sticking	0.5–1.0	X	X	X		X			X	X	
Coffee whiteners	Dispersant/emulsifier	0.4–1.0	X	X	X		X	X				
Chocolate	Quality improver, viscosity reducer, antiblooming	0.2–0.5	X	X	X		X	X				
Chewing gum	Quality improver, antisticking	0.2–1.0	X	X	X				X	X		
Soy/Worcester sauces	Quality improver	0.5–1.0	X	X	X				X	X		
Chocolate milk	Initial gloss, retards loss of gloss and fat bloom	0.8	X	X								
Coatings (non-standard) or confectionery	Usage depends on melting point of fat	0.8–1.0										X
Coating, panned	Reduces panning time, modifying sugar crystals	0.2										X
Noodles	Antistaling, improves extrusion and quality	0.5	X	X			X				X	
Prepared mix	Whipping agent, shelf-life improver	0.5	X	X	X		X				X	
Potatoes, dehydrated	Increases ease of rehydration, improves palatability	0.5–1	X	X	X	X		X				
Macaroni, spaghetti	Quality improver	0.5	X	X	X							
Marine paste products	Slippery agent	0.1–1	X	X	X							
Fish/sausage	Quality improver	0.1–0.5	X	X			X					X
Frozen fish paste	Quality improver	0.1–0.2	X	X	X							X
Instant food	Quality improver	0.5–3	X		X		X					X
Tofu	Defoaming and quality improver	0.5–0.8	X				X					X

Application	Function	Level (%)
Doughnuts, cake type	Volume, texture and shelf-life improver	8–10[2]
Doughnuts, yeast raised	Optimum fat absorption, retards firming, condition	10–14[2]
Dietary supplements	Emulsion stabilisation	0.1–0.2
Flavours	Improves dispersibility or solubility	2 and up
Ice-cream	Coalescent agent, provides dryness and overrun	0.2–0.3
Icings and fillings	Texture, volume, stability	1–4[2]
Shortening	Emulsion and antispattering	0.2–0.6
Margarine	Emulsion stability and palatability	0.5–5.0
Dressings	O/W emulsifier	0.1–0.5
Peanut butter	Inhibits oil separation and improves palatability	0.5–2.5
Pet foods	Retards firming, aids in extruding and inhibits separation	1–2.5
Pickles	Flavour dispersant	0.05
Processed food (defoaming)	Antifoaming agent	0.02–0.05
Salt	Controls crystal size	100 ppm
Sherbet	Increases whip, smooth texture, improve quality	0.05–0.1
Starch jellies	Retards starch crystallisation	0.25–0.5
Syrups	Stability	0.05
Whipped toppings	Aeration, dryness, body and texture	0.4–1.0
Yeast	Rehydration agent, improve shelf-life, defoamer	
Chinese steamed bread	Antistaling, quality improver	0–2
Cereals	Complexes starch, reduces sticking and clumping	2–4
Infant formula	Emulsification, provides body	0.1–0.4
Pet food	Extrusion aid, prevents sticking and clumping, improves shape	1–2.5
Diet spreads	Provides emulsion stability and water-holding capacity	1–5[2]
Calf milk replacer	Emulsion stability and palatability	0.2–0.5[2]

[1]These are suggested levels only. They could vary for each application and from country to country depending on regulations. In certain countries some suggested emulsifiers may not be approved.

[2]% of shortening.

Fatty acid soap: used in whipped topping, shortening and cake.

proportion of two fatty acid units to one glycerine unit. A diglyceride molecule may result from the combination of the units or the splitting of one fatty acid unit from a triglyceride during fat breakdown or hydrolysis.

Dispersion The distribution of finely divided solid particles in a liquid phase to produce a system of very high solid/liquid interfacial area.

Dispersion forces Weak interatomic or intermolecular forces common to all materials; generally attractive for materials in the ground state, although they can have a net repulsive effect in some solid/liquid systems.

Emulsifier A substance that at low concentration adsorbs at some or all of the interfaces in the system and significantly changes the amount of work required to expand those interfaces. The word 'surfactant' has been condensed from the term surface active agent. These substances have an *amphiphilic* nature. That is to say, they have chemical affinity to both lipid and aqueous phases. Because of this property, the molecules become oriented along the surfaces or interfaces of these normally immiscible substances. In food products the interface may be between two liquids, a liquid and a gas or a liquid and a solid. Molecules of such surfactants typically contain a hydrocarbon chain and a polar group. The hydrocarbon chain has affinity for lipids while the polar group has affinity to water or aqueous solutions. An emulsifier's impact on the interfacial tension at phase boundaries is dependent upon the polarity and solubility of the components in its chemical structure. Surfactants which have a high ratio of hydrocarbon groups to polar groups are lipophilic in nature and tend to be at least partially oil soluble. When a surfactant contains a high ratio of polar groups as compared to hydrocarbon groups, it will be hydrophilic in nature and tend to be at least partially water soluble.

Emulsion An emulsion is a two-phase system consisting of two incompletely miscible liquids, the one being dispersed as finite globules in the other. The dispersed, discontinuous, or internal phase is the liquid that is broken up into globules. The surrounding liquid is known as the continuous or external phase. Dispersions of other immiscible substances may also be made and may result in a foam (liquid or gas) or a suspension (liquid and solid). The characteristics of emulsions are not solely influenced by the choice of emulsifier(s), but also by such factors as surface-to-volume ratios, agitation, heating and cooling rates, entrainment of air and several other factors. Moreover, although most emulsions do not require a specific order of ingredient addition, the rate of addition is commonly critical to proper distribution and wetting of the emulsifier. Improved equipment designs have helped improve the consistency and quality of most common emulsions.

Ester An alcohol with one or more fatty acids attached. The most commonly found ester is a triglyceride, where an alcohol (glycerol) has three fatty acids attached. Monoglycerides (glycerol with one fatty acid attached) and diglycerides (glycerol with two fatty acids attached) are also esters.

Esterification The process of combining by chemical reaction an alcohol and an acid. The new product is an ester; water is a by-product of this reaction. A natural fat is a special type of ester made from glycerol (an alcohol) and fatty acids.

Foam booster A surfactant that will increase the amount of foam or its persistence in a given system.

Foam inhibitor A surfactant that will retard or prevent the formation of foam in a given system.

Flocculation The (often) reversible aggregation of drops or particles in which interfacial forces allow the close approach or touching of individual units. The separate identity of each unit is maintained.

Hydrophile–lipophile balance (HLB) An empirical method for quantifying the surface activity of a species based on its molecular constitution.

Hydrophilic ('water loving') A descriptive term indicating a tendency on the part of a species to interact strongly with water.

Hydrophobic ('water hating') The opposite of hydrophilic, having little energetically favourable interaction with water – generally indicating the same characteristics as lipophilic.

Interface The boundary between two immiscible phases. The phases may be solids, liquids, or vapours, although there cannot be an interface between two vapour phases. Mathematically, the interface may be described as an infinitely thin line or plane separating the bulk phases at which there will be a sharp transition in properties from those of one phase to those of the other, although in fact it will consist of a region of at least one molecular thickness, but often extending over longer distances.

Interfacial free energy The minimum amount of work required to create the interface or to expand it by unit area. When a surface tension measurement is performed on a liquid we are measuring the interfacial free energy per unit area of the boundary between the liquid and the air above it.

Interfacial tension The property of a liquid–liquid interface exhibiting the characteristics of a thin elastic membrane acting along the interface in such a way as to reduce the total interfacial area by an apparent contraction process. Thermodynamically, the interfacial excess free energy resulting from an imbalance of forces acting upon molecules of each phase at or near the interface.

Lipophilic ('fat loving') Materials that have a high affinity for organic solvents.

Lipophobic ('fat hating') The opposite of lipophilic, that is, materials preferring to be in a more polar or aqueous medium.

Lyophilic   A general term applied to a specific solute and solvent system, indicating the solubility relationship between the two.

Lyophobic   The opposite of lyophilic.

Micelles   Aggregated units composed of a number of molecules of a surface active material, formed as a result of the thermodynamics of the interactions between the solvent and lyophobic (or hydrophobic) portions of the molecule.

Monoglycerides   A chemical compound formed by the combination of one fatty acid unit with one glycerol unit. It may result from the combination of fatty acids and glycerol or from the partial hydrolysis of di- and triglycerides. Monoglycerides contain two types of chemical groups, one tending towards fat solubility and the other towards water solubility. The addition of monoglycerides to an oil or shortening tends to lower the smoke point of the oil.

Ostwald ripening   A process whereby large droplets will form at the expense of smaller droplets because the two phases forming the emulsion are not totally immiscible and there are differences in droplet size with the emulsion.

Solubilisation   The act of making a normally insoluble material soluble in a given medium.

Surface   An interface where one phase is a gas, usually air.

Phase inversion   Gradual increase in emulsion viscosity due to excess addition of a given phase until a critical concentration is reached. If more of that same phase is added exceeding the critical concentration, the emulsion will invert. (The discontinuous phase will become the continuous phase.)

Triglyceride   A chemical ester molecule formed by the reaction of one unit of glycerol with three units of fatty acids. A typical fat or oil will be a physical mixture of many different triglycerides.

Acknowledgement

The author extends his sincere appreciation to Mrs Brenda White for her endless effort in typing this manuscript.

Further reading

Arbuckle, W. S. (1986) *Ice Cream* (4th edn). Van Nostrand Reinhold, New York, p. 483.

Artz, W. E. (1990) Emulsifiers. In *Food Additives* (eds Branen, A. L., Davidson, P. M. and Salminen, S.). Marcel Dekker, New York, p. 736.

Becher, P. (ed.) (1983) *Encyclopaedia of Emulsion Technology*, Vol. 1: *Basic Theory*, Marcel Dekker, New York, p. 725.

Becher, P. (ed.) (1985) *Encyclopaedia of Emulsion Technology*, Vol. 2: *Applications*. Marcel Dekker, New York, p. 521.

Code of Federal Regulations (1986) *Food and Drugs*. Title 21. Parts 170–199 – Office of the Federal Register, National Archives and Records Administration, U.S. Government Printing Office, Washington, D.C.

Codex Alimentarius Commission (1979) *Guide to the Safe Use of Food Additives*. Second series. Pub. Food and Agriculture Organisation of the United Nations, World Health Organisation, Rome.

Council of European Communities (1985) Council Directive No. 85/6 on the approximation of the Laws of the member states relating to emulsifiers, stabilizers, thickeners and gelling agents for use in foodstuffs. Off. J. Eur. Communities 28, No. L2/21.

EFEMA (European Food Emulsifier Manufacturers Association) (1976) *Monographs for Emulsifiers for Foods*. Secretary EFEMA, Brussels, Belgium.

FAO/WHO (1979) *Guide to the Safe Use of Food Additives*. Issued by the Secretary of the Joint FAO/WHO Standards Programme, FAO, Rome.

Food Chemicals Codex (1981) (3rd edition). Committee on Codex Specifications, National Academy Press, Washington, D.C.

Food Emulsifiers, Chemistry, Technology, Functional Properties and Applications in Developments in Food Science (1980) (eds G. Charalambous and G. Doxastakis). Elsevier, New York, p. 549.

Friberg, S. (ed.) (1976) *Food Emulsions*. Marcel Dekker, New York, p. 480.

Griffin, W. C. (1965) Emulsions. In *Encyclopedia of Chemical Technology*, Vol. 8 (2nd edn). John Wiley, pp. 117–154.

Griffin, W. C. and Lynch, M. J. (1968) Surface Active Agents. In *Handbook of Food Additives* (ed. Furia, T. E.). The Chemical Rubber Co., Cleveland, Ohio, p. 771.

Kamel, B. S. (1984) Comparison of the Baker Compressimeter and the Instron in measuring firmness of bread containing various surfactants. *Cereal Food World* 29 (2), 159.

Knightly, W. H. (1988) Surfactants in baked foods: current practice and future trends. *Cereal Food World* 33 (5), 405.

Lynch, M.J., Griffin, W.C. (1974) Food emulsions. In *Emulsions and Emulsion Technology* (ed. Lissant, K. J.), Surfactant Science Ser., Vol. 6, Part 1. Marcel Dekker, New York.

Myers D. (1988) *Surfactant Science and Technology*. VCH Publishers, New York, p. 351.

Products for Foods (1976) Atkemix Inc., P.O. Box 1085, Brantford, Ontario, Canada, N3T 5T2, p. 20.

Rosen, M. J. (1989) *Surfactants and Interfacial Phenomena* (2nd edition). John Wiley and Sons, New York, p. 431.

Surfactants (1983). A comprehensive Guide published and edited in Japan by Kao Corporation, p. 500.

The HLB System (1976) Published by ICI Americas, Wilmington, Del., p. 19.

9 Bulking agents

T. E. LUALLEN

9.1 Introduction

The 1980s saw a trend towards the launching of new reduced-calorie or low-calorie food products. The calorie- and health-conscious consumer was ready and many food manufacturers responded.

Two approaches have been used to develop these new reduced-calorie foods: (1) portion control and (2) the reduction or elimination of high-calorie ingredients – primarily sweeteners and fats. The concept of portion control has worked within certain market segments. However, the reduction of calorie-contributing ingredients has received most attention. The food industry has invested a significant research effort into effective means of reducing fat content, lowering the level of sweetener solids and other nutritive carbohydrates.

Major advancements have been made on low- or non-caloric sweeteners. However, the majority of these alternative sweeteners are used at substantially lower levels than the original sweetener. This loss in sweetener solids has become a major problem in reduced-calorie products. It is the objective of this chapter to provide an overview of the products available for replacement of these calorie contributors and briefly describe their utilisation in various applications. An even more recent development is the surge in product development for reduced- or low-fat products. This creates the need for a whole new class of bulking agents as fat substitutes and fat replacers.

9.2 What are bulking agents?

To better clarify bulking agents as utilised today, the current trend in food technology focuses on the terms functional or non-functional bulking agents.

Ingredients added only to contribute bulk – over and above what is contributed by the regular formula of ingredients – should be classified 'non-functional'. Being basically a filler or extender, it produces no effect within the food matrix. There are relatively few situations where a bulking agent fits this description. Should, however, the bulking agent contribute to the control of certain factors it should be classified as a 'functional' bulking agent. Some of these functionalities are listed in Table 9.1.

Table 9.1 Effects of bulking agents (functional)

Sweetness	*Flavours*
– neutralise	– enhance
– reduce	– dispersant
– carrier	– carrier
Stabiliser	*Texture*
– moisture	– gels
– shelf-life	– pulp
– freshness	– grainy
Colours	*Body*
– dispersant	– chewiness
– carrier	– crispness
Mouthfeel	
– gummy	
– smoothness	

Trying to utilise bulking agents for more than 'bulk' factor has been a problem. In many foods, the ingredients utilised for bulking agents have frequently been described as bodying agents. This terminology puts a totally different view on the purpose of ingredients added as bulking agents. Bodying agents are commonly associated with textural qualities, i.e. mouthfeel or chewiness. Therefore, not all ingredients referred to as 'bodying agents' may be 'bulking agents'.

Bulking agents have also been classified as soluble or insoluble (Beereboom, 1979). These terms may still contribute to part of their definition, but not in total.

Traditional bulking agents were limited in design. They were considered inert, non-digestible and dry fillers. Today, a demand for functionality and calorie control has warranted totally different concepts for bulking agents to be considered as ingredients.

Sweetener solids replacement referred, in the past, to the reduction of sucrose. Today, technologists must also include regular corn syrups and high-fructose corn syrups. High-fructose corn syrup has essentially replaced sucrose in soft drinks. In other foods such as desserts, shelf-stable and frozen foods, high-fructose sweeteners have become the major sweetener ingredients.

Sweetener replacers (including polyols) are being sold at an annual rate of approximately 600 million pounds (Anderson, 1989). High-intensity sweeteners now account for approximately 13% (sweetness basis) of all sweeteners sold. It is estimated that within the next five years, approximately

50 million metric tonnes of sweetener replacers will replace approximately 300 million metric tonnes of sugar (*ibid.*). This tremendous change in sweetener solids will demand replacement ingredients (bulking agents) having the functional qualities of the removed sugar.

9.3 Traditional bulking agents

Fibres have been used frequently as bulking agents. These include the complex plant carbohydrates often referred to as 'dietary fibre', and forms of cellulose. Other terms that have become possibly more important – yet not recognised as a standard of quality – are soluble and insoluble fibres. There are many fibre sources. They can differ in dietary fibre, soluble fibre, insoluble fibre (Table 9.2), and in physical characteristics such as particle size.

Table 9.2 Fibre in cereals and legumes[1]

Fibre source	Total dietary fibre (% TDF)	Soluble fibre	Insoluble fibre
Cereals, legumes			
Field peas	85.0	14.4	70.6
Soybean bran	45–55	17.5	37.5
Barley bran	65.0	3.0	62.0
Corn bran	92	0.7–0.8	91
Oat bran	22.2	10.5	11.7
Rice bran	29–34	9.0	6–20
Rye bran	30–35	3–5.0	30.0
Wheat bran	42	3	39
Lentils, whole	20.3	—	—
White beans, whole	25.0	—	—
Pinto beans, whole	27.2	—	—
Sugar beet root	75–81	22–25	56–59

[1]McCormick (1988)

Fibre is used as a bulking agent for two primary reasons: (1) bulk and (2) calorie reduction. Excellent examples of how fibre is utilised as a bulking agent are in two commercial products: Nutrifat and Nutrifat FC. These products represent blends of soluble fibre and dextrins for use as fat substitutes. Nutrifat FC also contains some protein and is recommended for frozen desserts.

Many claims have been made pertaining to health benefits of fibre in the diet, but no attempt to judge these claims will be made in this review. Recommended fibre content per serving is shown in Table 9.3.

Maltodextrins are carbohydrates that have a lower molecular weight than

Table 9.3 Fibre in a serving of food

I. USA (FDA)
   3 g TDF – Good source
II. Canada (Health Protection Branch Report 5 February
   1988)
   2 g TDF – Moderate source
   4 g TDF – High source
   6 g TDF – Very high source

starch. They are considered Generally Recognised As Safe (GRAS) and by definition require a dextrose equivalence (DE) less than twenty (20). Dextrose equivalence represents the amount of reducing sugars present calculated as dextrose and expressed as a percentage.

During the late 1980s, several new maltodextrins were introduced. First came those classified as 'Low bulk density' (LBD) maltodextrins. These have a much lower apparent specific gravity compared to conventional maltodextrins. Low bulk density maltodextrins have apparent specific gravities as low as 0.05 g ml^{-1} (Table 9.4). These products can be produced by conventional

Table 9.4 Low bulk density maltodextrins[1]

Typical g ml^{-1}	Typical lb ft^{-3}	Typical DE (Dextrose equivalent)
0.01–0.10	3–6	5
0.21–0.30	13–19	5
0.05–0.10	3–6	10
0.21–0.30	13–19	10
0.14–0.19	8–12	20

[1]A. E. Staley Mfg. Co.

agglomeration techniques or by the use of blowing agents which produce hollow spheres, and typically have very high dispersion and solubility characteristics. They are very useful to prevent stratification of other dry ingredients in dry mixes, and because of their low caloric density they can effect direct caloric reductions.

A second maltodextrin generation has evolved from hydrolysates from carbohydrate sources other than corn. A product based on potato starch – Pasellii SA-2 (Avebe America, Inc.) – is listed as an enzyme-hydrolysed maltodextrin. It has also been reported that maltodextrins can be prepared from milled rice flour using a heat-stable α-amylase preparation (Griffin and Brooks, 1989). Considering the variable properties, composition and textural characteristics available from different native starches, maltodextrins derived from each could impart significant effects within a food matrix. Maltodextrins derived from these starches may be new bulking agents for the 1990s.

9.3.1 *Bulking agents as sugar replacers and flavouring agents*

Many of the new sweetening agents available to replace sugar lack many functional properties of sugar. The many high-intensity sweeteners available account only for the sweetness factor of sugar. However, as these intense sweeteners replace more and more sucrose, the need for functional bulking agents will expand. Table 9.5 contains a partial listing and status of some of the new high-intensity sweeteners.

The sugar replacers listed in Table 9.5 contribute some degree of sweetness. For those food technologists formulating with sugar, who want to maintain the functional attributes and yet reduce or eliminate the sweetness, 'Lactisole' (Refined Sugars, Inc.) may be an appropriate alternative. Listed as a flavour agent, it balances flavour perception by reducing the perceived sweetness. This ingredient allows the use of high levels of sugars and polyols to achieve desired functionality without overwhelming sweetness. Food technologists now have another opportunity to utilise the functional properties of carbohydrates – texture, body, dispersion and control the sweetness (Table 9.6). Lactisole has not yet been approved for general use.

Polydextrose (Beereboom, 1979) is a relatively bland-tasting, water-soluble polymer of randomly branched glucose residues. Polydextrose (Murray, 1988) contains minor amounts of bound sorbitol and citric acid. It has been approved for a wide variety of applications: baked sweet goods, confections, salad dressings, some desserts, chewing gum and mayonnaise. Some of its characteristics are listed in Table 9.7. Some of the energy-reduction potential of polydextrose is shown in Table 9.8.

9.3.2 *Sugar alcohols*

Palatinit (Higginbotham, 1983) is prepared by the enzymic conversion of sucrose to isomaltulose and the subsequent hydrogenation to Palatinit. Some of its characterisics are shown in Table 9.9.

Products such as lactitol and maltitol are also being evaluated as reduced-energy ingredients (Anderson, 1989). There is some controversy as to their actual energy value. A value of $2\,\mathrm{kcal\,g^{-1}}$ is assumed by many scientists. It is too early to predict their use in future applications.

9.3.3 *Polyols*

A family of products very closely related to sugar is polyols, which are sugar alcohols. They possess many physical properties similar to sugars, especially liquid sugars. Sorbitol, xylitol and mannitol can be used in sugar-free

Table 9.5 High-intensity sweeteners and flavour agents

Sweetener	Description	Sweetness intensity ($\times$ sucrose)	Regulatory status
Saccharin	Synthetic, non-caloric, heat stable, bitter after-taste at high concentration	300	Permitted in US Congressional Action; approved in over 80 countries
Cyclamate	Synthetic, non-caloric, heat stable, no after-taste	30	Not approved in US; approved in 40 countries
Acesulfame-K	Synthetic, non-caloric, heat stable, synergistic effect in sweetener blends	200	Approved in US (7/88) for some uses (dry goods); approved in 20 countries; approval *to be sought* for use in liquids and baked goods
Aspartame	Dipeptide, low-calorie nutritive; not stable to prolonged heating	180–200	Approved in US for certain uses (including dry goods, beverages); petition filed (9/87) in US for use in baked goods and baking mixes
Alitame	Dipeptide, low-calorie nutritive; clean taste, no after-taste; synergistic effect in sweetener blends; relatively heat stable	2000	Petition filed (8/86) in US to include baked goods
Sucralose	Synthetic, non-caloric; Chloro-derivative of sucrose; heat stable	600	Petition filed (1987) in US to include baked goods and baking mixes
L-Sugars	Synthetic, non-caloric; optical isomer of common, natural sugars; not absorbed by humans; identical physical and chemical properties; no after-taste	1	Not approved in US; petition planned by toxicology studies initiated in 1987; to be completed by 1990
Thaumatin (Talin)	Protein, low-caloric nutritive; delayed sweetness; liquorice-like after-taste; genetically engineered product recently reported	2000	Not approved in US; available in UK and , Japan, GRAS as flavour adjunct; petition planned for about 1989
PS-99	Dipeptide, low-caloric nutritive	1800–2200	Not approved; toxicology testing initiated; approval at least 4–6 years away

Table 9.6 Usage rate for Lactisole[1]

Application level (ppm)	Effect	Examples
Up to 50	Flavour improvement	Fruit jams, fondants, extruded cereal products
50–100	Sweetness modification	Baked goods, liqueurs, reformed foods and confectionery
100–150	Loss of sweetness	Baked goods

[1]Refined Sugars, Inc.

Table 9.7 Polydextrose[1]

Advantages
Versatility; it can be used in a wide variety of products
Non-cariogenic
1 kcal g^{-1}
Functional as a tenderiser in baked goods
Good humectant and moisture stabiliser
Flow properties are Newtonian
Melts at 130°C and cools to a hard glass, but does not crystallise
Good fat-like textural qualities in frozen desserts

Disadvantages
Average daily intake (ADI) limit of 90 g
If greater than 15 g per serving, a warning label is required stating that this product
    may have a laxative effect
At high usage levels, off-flavours can be noted
Usage levels are limited in products which require the functionality of sucrose
Does not crystallise

[1]Anderson (1989)

Table 9.8 Average energy-reduction potential of polydextrose in a variety of products[1]

Product	Energy reduction	Use level (by weight)
Ice-cream	50%	15%
Baked goods	33%	N/A
Pudding, instant	70%	50%
Confectionery	33%	N/A
Yoghurt	N/A	14%
Jams	N/A	25%

[1]Pfizer, Inc., USA.

applications (non-cariogenic). However, they do not significantly reduce the energy value of the food products.

9.3.4 *Bulking agents as fat substitutes and extenders*

Bulking agents utilised as fat substitutes can be divided into two categories.

Table 9.9 Palatinit[1]

Advantages
Compared to other sugar alcohols, it has a higher tendency to crystallise, lower hygroscopicity,
  and no cooling effect in the mouth
Non-cariogenic
Functions well in low-moisture systems
Permits the use of standard processes and equipment in the manufacture of hard-boiled candy
Outperforms all other sugar alcohols in candy manufacture
Resistant to chemical change such as hydrolysis or enzyme activity
$2 \, \text{kcal} \, \text{g}^{-1}$
Approved for use in several European countries

Disadvantages
Because it is only half as soluble as sucrose, crystallisation occurs in products such as jams
  and jellies where the solids are greater than 25%
Some intolerance to Palatinit by a few individuals
Not approved for use in US

[1]Strater (1988).

Those substitutes which only partially mimic or duplicate fat (lipid) properties have been referred to as fat-sparing. They lack functionality. In a second
category are those bulking agents that possess both characteristics and
functionality – a total fat substitute. Usually the greater the percentage of
fat in a food system, the more difficult it is to obtain all characteristics and
functional properties from a single substitute.

  Water-soluble polymers are often used as partial or total fat substitutes.
They can include tree and shrub exudates, seed and seaweed extracts,
cellulose ethers, guar gum, xanthan gum and their derivatives (Higginbotham, 1983). Traditionally, water-soluble polymers have been used to
stabilise the water-phase portion of food systems. Others are used to impart
body, texture and viscosity. In many instances, modified starches are utilised
with gums to reduce cost and impart or enhance certain textures.

  Modification techniques that enhance starches as potential fat substitutes
are being researched extensively. These modified starches act as fat substitutes in some instances and as fat extenders in others. The origin of the
base starch and type of modification enhance the functionality of one starch
compared to another. Hydrolysed potato starch, at the present time, seems
to provide some of the best starches for fat substitution. The hydrolysis can
be accomplished by using enzymes or acids to cleave some of the molecular
bonds of the starch molecule.

  Corn-derived starches, specifically waxy maize, when substituted with 1-
octynyl succinic anhydride, imparts excellent fat-binding properties. These
starches are not fat substitutes, but can be utilised to extend or enhance fat
properties in some low-fat products. Combining the substitution and hydrolysis treatments accentuates these properties and also imparts emulsifying characteristics. A partial listing of some commercially available starch-
based fat substitutes appears in Table 9.10.

  Some non-nutritive oil-based fat replacers are also being developed. A

Table 9.10 Examples of industrially supplied modified starches as substitutes/replacers/emulsifiers

Product	Carbohydrate base
STA-SLIM™ 143[1]	Potato
PURITY GUM™ BE[2]	Waxy maize
N-OIL™[1]	Tapioca
STA-MIST™ 515[1]	Waxy maize
Crystal Gum[2]	Tapioca
Pasellii SA[3]	Potato
N-ZORBIT™[2]	Tapioca
STA-MIST™ 7415[1]	Waxy maize
N-FLATE[2]	Starch/guar gum
Instant N-OIL[2]	Tapioca
Instant STA-SLIM™[2]	Potato
PURITY GUM 539™[2]	Waxy maize

[1]A. E. Staley Mfg. Co.
[2]National Starch & Chem. Co.
[3]Avebe America, Inc.

Table 9.11 Developmental fat replacers

Olestra®[1] is a highly substituted sucrose polyester. It is regarded as non-digestible and considered non-caloric. It is reported to behave nearly identically to triglycerides. However, it is not an approved ingredient at this time.

Simplesse[2] is a reduced-sized natural protein based on milk protein and/or egg whites.

Prolestra[3] is a sucrose polyester plus protein. It is said to be able to replace up to 10% of the fat in certain food systems.

Polysiloxanes[4] are polyphenyl-methylsiloxene (PPMS) or polydimethylsiloxane (PDMS). This is not an approved product; however, it is in the animal feeding stage.

Ethyoxylated monoglycerides[5] are lipids esterified with ethylene oxide. They have been approved in the United States, but not in Japan.

Alkoxylated fatty acids[6] (alcohols or esterified polypoxylated glycerol) are glycerol, fatty alcohols and fatty acids esterified with ethylene and/or propylene oxides.

Substituted malonyl esters[7] are long-chain alkyl chlorides of malonic acid or esters. They possess good thermal properties; however, the raw material costs are excessive.

[1]Proctor & Gamble;
[2]NutraSweet;
[3]Aries (1989);
[4]Hashim (1989);
[5]Egan *et al.* (1969) and MacDonald *et al.* (1964);
[6]ARCO;
[7]Frito Lay.

partial list is shown in Table 9.11, but these are, however, beyond the subject of this review.

9.3.5 *Utilisation of bulking agents*

The utilisation of bulking agents may be best understood by first reviewing some of the accepted and regulated standards for energy levels. An accepted level, for instance, in frozen entrées or dinners is 300 calories. The United States Code of Federal Regulations, Title 21, Section 105.66 states these standards: Foods which have one-third (1/3) fewer calories relative to their full calorie counterparts are listed as 'reduced calorie'. Foods that have been reduced in calories to not more than 0.4 kcal g^{-1} and limited to not more than forty (40) calories per serving shall be labelled 'low calorie'.

9.3.6 Dressings, sauces and mayonnaises

The primary energy contributor in these foods is the fat or oil. Sweeteners and starches (flour) also contribute some energy, but fat is the primary calorie source. Today, modified starches and maltodextrins are being suggested as partial and, in some instances, total replacers for oils used in dressings.

Ten years ago, products that did not contain sufficient ingredient quantities for the standards of mayonnaise had to be labelled as 'imitation'. Today similar formulations are classified as 'reduced calorie'. One important thing to remember is that most bulking agents are *not* water-phase stabilisers. Therefore, other hydrocolloids designed for water-phase stabilisation are necessary. Polydextrose can be utilised to assist with textural properties that many low-calorie dressings lack. A closer sensory match of the full calorie product may be possible with polydextrose as an additive.

9.3.7 Desserts

Another large market potential for reduced-energy foods is desserts. Imitation ice-cream and frozen novelties are examples. Maltodextrins, modified starches, specially prepared proteins, and water-soluble polymers are bulking agents that can replace part of the fat, while meeting many of the full energy product characteristics. Polydextrose, low bulk density maltodextrins, and gums are typical bulking agents for dry mix desserts. Body, texture and dispersibility are the functional characteristics desired.

9.3.8 Confectionery

Confections form a huge consumer market with a very large potential for bulking agents. However, polydextrose at present is the only commercially acceptable one in use. Polyols (mannitol, sorbitol, xylitol) have assisted with sucrose replacement (non-cariogenic), but do not offer calorie reduction.

9.3.9 Dry mixes

As previously mentioned in dry mix desserts, bulking agents do have potential in this market segment. They may be purely non-functional, as previously discussed, or they may be flavour or oil carriers as well as providing bulk and dispersibility. Water-soluble polymers, low bulk density maltodextrins, modified starches, and high-intensity sweeteners offer a wide array of product potential. Energy reduction can be done by weight or volume in these mixes.

9.3.10 *Bakery products*

In many baked products, portion control has been a widely used method for energy reductions. However, dietary fibre and/or α-cellulose addition to breads, etc., have added bulk with energy reduction. Polydextrose in a few bakery products has also been utilised, primarily in cream-filled cookies.

Another segment of the bakery area is fillings. Cream fillings, as previously mentioned, are one type. Fruit fillings are a successful market for bulking agents that can replace sucrose. Some major US companies are supplying reduced-energy bakery fillings.

9.3.11 *Dairy*

Excluding the previously discussed ice-cream and dessert-type products, this market segment still offers a wide range of high-fat food products available for reformulation with bulking agents. Margarine, for example, ranges from 40 to 80% fat. The yoghurt, cream and cheese industries are introducing lower fat content products. Water-soluble polymers, modified starches and maltodextrins are acceptable bulking agents. Mouthfeel, texture and stability are primary product characteristics to be matched when compared to the full-energy product.

9.4 Conclusion

As can be seen, many currently available ingredients fall short of contributing the functional properties, health characteristics and consumer appeal necessary for the sweetener replacements and fat substitutes desired in today's food products. Therefore, new bulking agents which contribute the properties necessary to match full-calorie foods are required for the future.

The 1990s should bring many new and exciting ingredients for reduced-energy foods. Several high-intensity sweeteners are on the horizon, but there is still a shortage of functional fat and sucrose substitutes. Consumer safety being a primary concern, regulatory agencies may limit approvals for new ingredients. Therefore, a need for functional bulking agents would be a likely alternative.

The ideal fat or sweetener substitute of the future will need to possess these characteristics:

Ideal characteristics

Fat substitute	*Sweetener replacer*
Energy value of less than $4\,kcal\,g^{-1}$	Energy value of less than $2\,kcal\,g^{-1}$
Derived from an acceptable source	Derived from an acceptable source
Toxicologically safe in humans	Toxicologically safe in humans
Free of any osmotic diarrhoea or anal leakage	Free of any osmotic diarrhoea or anal leakage
Unrestricted usage levels	Unrestricted usage levels
Unobjectionable taste and colour, bland and colourless	Unobjectionable taste or colour, bland and colourless
Comparable price to fats and oils	Comparable price to sucrose
Functionally similar to fats and oils in all applications	Sweet as sucrose
Withstands deep fat frying temperatures for days	Functionality similar to sugar in all applications
Fat-like mouthfeel	
Adjustable melt-point tailored around specific applications	

These bulking agents will help optimise the use of current reduced-energy ingredients. They will contribute those qualities necessary to match more closely the functionality and sensory properties of fats and sugars.

References

A. E. Staley Mfg. Co., P.O. Box 251, Decatur, IL 61525. Technical Literature.

Anderson, K. (1989) Ingredients for reduced calorie foods. Abstract. *International Symposium Foods for the '90's*, London.

Aries, R. S. (1989) Prolestra – a new sucrose polyester and protein composition. *JAOCS* **66** (4), 470.

Avebe America, Inc., Princeton Corporate Center, 4 Independence Way, Princeton, NJ 08540.

Beereboom, J. J. (1979) Low calorie bulking agents. *CRC. Crit. Rev. Food Sci. Nutri.* **11**, 401–413.

Egan, R. R. *et al.* (1969) *Ethoxylated Monoglycerides in Baked Goods*. US Patent No. 3,433,645; 18 May.

MacDonald, I. A., Egan, R. R. and Lampson, S. B. (1964) *Emulsifier for Frozen Confections*. US Patent No. 3,821,442; 20 June.

Hashim, S. A. (1989) Polysiloxanes: potential non-caloric fat substitutes. *JAOCS* **66** (4), 480.

Higginbotham, J. D. (1983) *Developments in Sweetness – 2* (eds Gremby, T. H., Parker, K. J. and Lindley, M. G. Applied Science Publishers, London, pp. 225–246.

McCormick, R. (1988) Function and Nutrition Guide Fiber Ingredient Selections, *Prepared Foods*, **157** (12) p. 83.

Murray, P. R. (1988) *Low Calorie Products* (eds Birch, G. G. and Lindley, M. G.). Elsevier Applied Science, London, pp. 83–100.

National Starch and Chemical Corporation, Fenderne Avenue, P.O. Box 6500, Bridgewater, NJ 08807, Product Publication, Bulletins No. 287, 500, 247, 457, 19483, 508, 262–A.

NutraSweet™ – Monsanto. Sophia Antipolis, Valbonne. Cedex, 06561, France.

Proctor & Gamble. One Procter & Gamble Plaza, Cincinnati, OH45202.

Refined Sugars, Inc., 1 Federal S. Yonkers, NY 10702. Technical Publication.

Strater, P. J. (1988) *Low Calorie Products* (eds Birch, G. G. and Lindley, M. G.). Elsevier Applied Science, London, pp. 63–82.

Further reading

d'Ostiowick, P. and de Roocker, A. (1972) *Preparation of Diholo Compounds*. US Patent No. 3,793,380; 21 November.

Fulcher, J. (1986) *Synthetic Cooking Oils containing Dicarboxylic Acid Esters*. US Patent No. 4,582,927; 15 April, Note 2.

Griffin, V. K. and Brooks, J. R. (1989) Production and size distribution of rice maltodextrins hydrolyzed from milled rice flour using heat-stable alpha-amylase. *Journal of Food Science* **54**, No. 1, 190.

HRA, Inc., 4510 W 89th St, Prairie Village, KS 66207–2282, Technical Publication.

Lineback, D. (1988) North Carolina State, presentation to Nat. Biscuit & Cracker, November.

Luallen, T. E. (1988) Structure, characteristics and uses of some typical carbohydrate food ingredients. *Cereal Foods World* **33**, No. 11, 924.

White, J. F. and Pollard, M. R. (1988) *Non-digestible Fat Substitutes of Low-caloric Value*. European Patent No. 254,547, 27 January.

Wolkstein, M. (1989) Nutrifat – a natural products fat replacer. *JAOCS* **66** (4) 481.

10 pH control agents

P. de la TEJA

10.1 Introduction

This chapter gives an overview of the most frequently used acidulants in the food industry. Tables describing the use of a typical acidulant (ascorbic acid) as a food-processing agent are included. The emphasis of the chapter has been given to processing and formulation with these pH-adjusting agents.

For easy reference, a list of tables in this chapter is given here.

Table 10.1 Properties of some common food acidulants
Table 10.2 Some natural acids of fruit and vegetables
Table 10.3 Effect of ascorbic acid and sodium ascorbate in cured meat colour
Table 10.4 Physicochemical properties of ascorbic acid
Table 10.5 Inhibitory effect of ascorbic acid on polyphenoloxidase in pears
Table 10.6 Effect of ascorbic acid on colour of sliced peaches
Table 10.7 Examples of fruit dipping solutions
Table 10.8 Effect of ascorbic acid and calcium on quality of frozen sliced apples
Table 10.9 Effect of ascorbic acid on canned sliced apples
Table 10.10 Effect of ascorbic acid on the mixing and kneading rate and bread quality
Table 10.11 Effect of ascorbic acid in white wine
Table 10.12 Examples of use of ascorbic acid in oils and fats
Table 10.13 Method of adding ascorbic acid to wine

Table 10.1 Properties of some common food acidulants[1]

	CFR[2]	Ionisation constant(s)/pK_a	Physical form	Melting point (°C)	Solubility (g (100 ml)$^{-1}$ water)	Hygro-scopicity	Taste characteristics	Empirical formula	Molecular weight	Density
Acetic acid	184.1005 (GRAS)	176×10^{-5}/4.75 at 25°C	Clear, colourless liquid	−8.5	Miscible	—	Tart and sour	$C_2H_4O_2$	60.05	d_{25}^{25} 1.049
Adipic acid	184.1009 (GRAS)	$K_1 = 3.71 \times 10^{-5}$/4.43 $K_2 = 3.87 \times 10^{-6}$/5.41 at 25°C	Crystalline powder	152	1.9 g at 20°C	Low level of hygroscopicity	Smooth lingering tartness; complements grape flavours	$C_6H_{10}O_4$	146.14	d_4^{25} 1.360
Citric acid	182.1003 (GRAS)	$K_1 = 7.10 \times 10^{-4}$/3.14 $K_2 = 1.68 \times 10^{-5}$/4.77 $K_3 = 6.4 \times 10^{-7}$/6.39 at 20°C	Crystalline powder	Anhydrous 153 Hydrated, 135–153	181 g at 25°C 208 g at 25°C	Moderately hygroscopic	Tart; delivers a 'burst' of tartness	$C_6H_8O_7$	192.12	1.665
Fumaric acid	172.350 (food additive)	$K_1 = 9.30 \times 10^{-4}$/3.03 $K_2 = 3.62 \times 10^{-5}$/4.44 at 18°C	White granules or crystalline powder	286	0.5 g at 20°C 9.8 g at 100°C	Non-hygroscopic	Tart; has an affinity for grape flavours	$C_4H_4O_4$	116.07	1.625
Glucono δ-lactone	184.1318 (GRAS)	1.99×10^{-4}/3.7 (for gluconic acid)	White crystalline powder	153	59 g at 25°C	Non-hygroscopic	Neutral taste with acidic aftertaste when hydrolysed	$C_6H_{10}O_6$	178.14	—

Name	CFR No.	Dissociation constants	Physical form	m.p. (°C)	Solubility	Hygroscopicity	Taste	Formula	Mol. wt	Density
Lactic acid	184.1061 (GRAS)	$1.374 \times 10^{-4}/3.86$ at 25°C	Liquid, also available in dry form	16.8	Very soluble	—	Acrid	$C_3H_6O_3$	90.08	—
Malic acid	184.1069 (GRAS)	$K_1 = 3.9 \times 10^{-4}/3.40$ $K_2 = 7.8 \times 10^{-6}/5.11$ at 25°C	Crystalline powder	132	62 g at 25°C	Non-hygroscopic	Smooth tartness	$C_4H_6O_5$	134.09	—
Phosphoric acid	182.1073 (GRAS)	$K_1 = 7.52 \times 10^{-3}/2.12$ $K_2 = 6.23 \times 10^{-8}/7.21$ $K_3 = 2.2 \times 10^{-13}/12.67$ K_1 and K_2 at 25°C K_3 at 18°C	Liquid		Very soluble in hot water	—	Acrid	H_3O_4P	98.00	1.8741 (100% sol.) 1.6850 (85% sol.) 1.3334 (50% sol.) 1.0523 (10% sol.)
Tartaric acid	184.1099 (GRAS)	$K_1 = 1.04 \times 10^{-3}/2.98$ $K_2 = 4.55 \times 10^{-5}/4.34$ at 25°C	Crystalline powder	168–170	147 g at 25°C	Non-hygroscopic	Extremely tart; augments fruit flavours, especially grape and lime	$C_4H_6O_6$	150.09	d_4^{20} 1.7598
Succinic acid	184.1091 (GRAS)	$K_1 = 6.5 \times 10^{-5}$ $K_2 = 2.3 \times 10^{-6}$	White crystalline powder	188			Tart	$C_4H_6O_4$	118.09	d_4^{15} 1.564
Succinic anhydride	172.892 (GRAS)	$K_1 = 6.5 \times 10^{-5}$ $K_2 = 2.3 \times 10^{-6}$	White crystals	118.3, 119[3]			Burning tart	$C_4H_4O_3$	100.07	d_4^{20} 1.503

[1] Reprinted with the permission of Food Technology, Allied-Signal, Inc.
[2] Title 21 of the Code of Federal Regulations (FDA, 1988).
[3] Solidification point.

Table 10.2 Some natural acids of fruits and vegetables[1]

Fruits

Apples	Malic, quinic, α-ketoglutaric, oxalacetic, citric, pyruvic, fumaric, lactic and succinic acids
Apricots	Malic and citric acids
Avocados	Tartaric acid
Bananas	Malic, citric, tartaric and traces of acetic and formic acids
Blackberries	Isocitric, malic, lactoisocitric, shikimic, quinic and traces of citric and oxalic acids
Blueberries	Citric, malic, glyceric, citramalic, glycolic, succinic, glucuronic, galacturonic, shikimic, quinic, glutamic and aspartic acids
Boysenberries	Citric, malic and isocitric acids
Cherries	Malic, citric, tartaric, succinic, quinic, shikimic, glyceric and glycolic acids
Cranberries	Citric, malic and benzoic acids
Currants	Citric, tartaric, malic and succinic acids
Elderberries	Citric, malic, shikimic and quinic acids
Figs	Citric, malic and acetic acids
Gooseberries	Citric, malic, shikimic and quinic acids
Grapefruit	Citric, tartaric, malic and oxalic acids
Grapes	Malic and tartaric (3:2), citric and oxalic acids
Lemons	Citric, malic, tartaric and oxalic acids (no isocitric acid)
Limes	Citric, malic, tartaric and oxalic acids
Orange Peel	Malic, citric and oxalic acids
Oranges	Citric, malic and oxalic acids
Peaches	Malic and citric acids
Pears	Malic, citric, tartaric and oxalic acids
Pineapples	Citric and malic acids
Plums	Malic, tartaric and oxalic acids
Quinces	Malic acid (no citric acid)
Strawberries	Citric, malic, shikimic, succinic, glyceric, glycolic and aspartic acids
Youngberries	Citric, malic and isocitric acids

Vegetables

Beans	Citric, malic and small amounts of succinic and fumaric acids
Broccoli	Malic and citric (3:2) and oxalic and succinic acids
Carrots	Malic, citric, isocitric, succinic and fumaric acids
Mushrooms	Lactarimic, cetostearic, fumaric and allantoic acids
Peas	Malic acid
Potatoes	Malic, citric, oxalic, phosphoric and pyroglutamic acids
Rhubarb	Malic, citric and oxalic acids
Tomatoes	Citric, malic, oxalic, succinic, glycolic, tartaric, phosphoric, hydrochloric, sulfuric, fumaric, galacturonic and pyrrolidinonecarboxylic acids

[1]Reprinted with permission of Allied-Signal, Inc.

Table 10.3 Effect of ascorbic acid and sodium ascorbate in cured meat colour

	Salting time		
Sample	3 hours	4 hours	5 hours
Control	Colour was almost developed	Colour was almost developed	Colour was completely developed
Ascorbic acid	Colour was almost developed	Colour was completely developed	Colour was completely developed
Sodium ascorbate	Colour was almost developed	Colour was completely developed	Colour was completely developed

Meat used: beef.
Salt 3%, sodium nitrite 0.02%, vitamin C 0.05% or sodium ascorbate 0.055%.
Salting process conditions: dry method, 5°C.

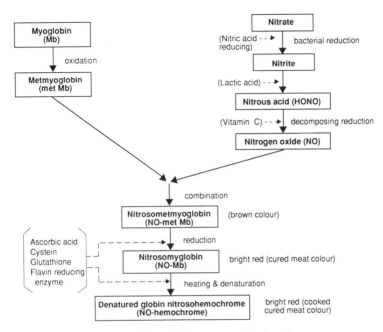

Figure 10.1 Colour development mechanism of meat.

Table 10.4 Physicochemical properties of ascorbic acid

	Ascorbic acid (Vitamin C)	Sodium ascorbate
Empirical formula	$C_6H_8O_6$	$C_6H_7NaO_6$
Molecular weight	176.13	198.11
Melting point	187–192°C	—
Properties	White or slightly yellow crystals or crystalline powder; odourless with an acid taste	White or very faintly yellow crystals or crystalline powder; odourless with a slightly salty taste
Solubility	1 g is soluble in 3 ml of water and in 50 ml of ethanol	1 g is soluble in 2 ml of water and scarcely soluble in ethanol
Quality of solution	pH = 3 (0.5% Aq. soln) pH = 2 (5% Aq. soln)	pH = 6.5–8.0 (2% Aq. soln)

Table 10.5 Inhibitory effect of ascorbic acid on polyphenoloxidase in pears (% inhibition)

	Level of inhibitor (mM)			
Inhibitory agent	0	0.05	0.5	5.0
Ascorbic acid	0	53.1	100.0	100.0
Erythorbic acid	0	53.6	100.0	100.0
Diethyldithiocarbonate	0	12.4	60.0	100.0
Phloroglucinol	0	7.0	26.8	52.7
Iodacetamide	0	0	0	0

Reaction mixture: 50 ml of 0.06 M catechol + 1 ml pear fermentation liquid + inhibiting agent, pH 6.2, 30°C

Table 10.6 Effect of ascorbic acid on colour of sliced peaches

Antioxidant or other additive	Amount added	Treatment	Period without discoloration	Colour change
Ascorbic acid	0.05%	None	60 days	Top syrup after opening
	0.05%	Vacuum fill	60 days	None
	0.05%	Heated at 49°C, 5 minutes	60 days	None
Peroxidase inhibitor	12 ppm	None	60 days	Faded
	12 ppm	Heated at 49°C, 5 minutes	14 days	Faded
Oxygen accepter	50 ppm	None	8 days	Surface discoloured
	50 ppm	Heated at 49°C, 5 minutes	8 days	Surface markedly discoloured
BHT	0.1%	None	7 days	Surface markedly discoloured
	0.1%	Heated at 49°C, 5 minutes	7 days	Surface markedly discoloured

Packing: Glass jar, 7 oz peaches in 4 oz syrup.
Storage condition: 0°C.

Table 10.7 Examples of fruit dipping solutions

Fruit	Acidulant			
	Ascorbic acid	Calcium salt	Citric acid	Malic acid
Apples	0.02–0.05	—	—	—
	0.01	0.1	—	—
	0.2–0.4	—	—	—
Oranges	5.0	—	—	—
	0.1	—	2.0	—
	0.5	0.5	—	—
	1.0	—	—	0.025

Table 10.8 Effect of ascorbic acid and calcium on quality of frozen sliced apples

Ascorbic acid (%)	Calcium (%)	Shear strength	Quality and appearance		
			Texture	Colour	Flavour
0.05	0.3	34	Hard, good	Slightly browned	Pleasant
0.2	0.3	54	Good	Good	Good
0.5	0.3	41	Soft	Good	Good
2.0	0.3	17	Soft	Good, bright yellow	Good
1.0	0.4	48	Very hard, good	Good	Good

Treated with 30% sucrose solution with ascorbic acid and $CaCl_2$. A few days after freezing, apples were thawed for 6 hours at room temperature and then evaluated.

Table 10.9 Effect of ascorbic acid on canned sliced apples

Ascorbic acid addition rate mg $(jar)^{-1}$	Total ascorbic acid content mg $(jar)^{-1}$	After 6 months storage at 24°C	
		Ascorbic acid content mg $(jar)^{-1}$	Appearance
0	3	5	Completely blackened
67	18	15	Slightly blackened
135	59	33	Normal
200	118	82	Normal

Apple slices were cooked 3–4 minutes and hot-packed in 1-pint glass jar with 50% syrup, then heated 10 minutes in boiling water.

Table 10.10 Effect of ascorbic acid on the mixing and kneading rate and bread quality

	Mixing and kneading rate (rpm)	Volume (ml)	Appearance score
Without added ascorbic acid	230	4130	86
Ascorbic acid added (40 ppm)	190	4150	87

Ingredients: flour, 100; water, q.s; sugar, 8.0; salt, 2.0; yeast, 3.0; yeast food, 0.5; lard, 2.7; surfactant, 0.25; potassium bromate, 0.006; potassium iodine, 0.001.

Table 10.11 Effect of ascorbic acid in white wine

	Amount ascorbic acid added (ppm)		
	0	120	240
Absorbance			
D_{430}	0.104 (Brown)	0.082	0.081
D_{530}	0.044 (Brown)	0.085	0.030
Residual ascorbic acid (%)	—	72.4	79.6
Sulphur content: Free	4.1	4.9	7.4
Total	9.4	14.6	16.4

Bottled wine was tested after 6 months' storage in a cool, dark place.

Table 10.12 Examples of use of ascorbic acid in oils and fats

Oils and fats and food containing them	Addition rate of ascorbic acid	Type and concentration of antioxidant used in combination with ascorbic acid
Lard	0.01–0.04%	Tocopherol
Lard	0.05%	0.1% of NDGA
Lard	0.1–0.15%	0.05% of NDGA + 2% of α-tocopherol
Lard	0.002–0.2%	10^{-3}–10^{-5}M PG, NDGA, BHA and BHT
Lard, goose fat	0.02%	0.1% of PG and BHA
Lard, suet chicken fat	0.02%	0.2% of d1-α-tocopherol and 0.02% of d1-γ-tocopherol
Cream	0.075%	0.009–0.012% of α-tocopherol

Table 10.13 Method of adding ascorbic acid to wine

Ascorbic acid addition (mg l^{-1})	Process at which ascorbic acid was added	Content of free sulphurous acid in wine (mg l^{-1})
50–100	Just before bottling	15–20
50	After fermentation period	25
50–70	Just after bottling	15–20

Further reading

Confidine, D. M. and Confidence, G. D. (1982) *Food and Food Production Encyclopedia*. Van Nostrand Reinhold, New York.

Dziezak, Judie D. (1990) Ingredients that do more than meet the acid test. *Food Technology* **44** (1), 76–83.

Food Acidulants. Allied-Signal, Inc.

Food Chemical Codex.

Merck Index (1983) 10th edition. Merck and Co. Inc., Rahway, NJ, USA.

Takeda Technical Literature, Takeda, Osaka, Tokyo, Japan:

(a) *Food Processing with Vitamin C for Beverages*, Series 1.
(b) *Food Processing with Vitamin C for Meat Products*, Series 2.
(c) *Food Processing with Vitamin C for Fruit Products*, Series 3.
(d) *Food Processing with Vitamin C for Flour Products*, Series 4.
(e) *Food Processing with Vitamin C for Oils and Fats*, Series 5.
(f) *Food Processing with Vitamin C for Beer and Wine*, Series 6.
(g) *Food Processing with Vitamin C for Vegetable Products*, Series 7.
(h) *Food Processing with Vitamin C for Dairy Products*, Series 8.
(i) *Food Processing with Vitamin C for Sea Foods*, Series 9.

11 Hydrocolloids

J. E. TRUDSO

11.1 Introduction

The purpose of this chapter is to present practical, applicable tools for people in product development, tools that will facilitate a screening and rough selection of hydrocolloids as possible candidates for fulfilling technological needs in processed food products.

Before these tools can be used, the technological needs must have been defined, and the chapter ends with an example showing this.

Since the chapter has intentionally been made as simple as possible, not all aspects of the properties of hydrocolloids are included. There are, indeed, a number of exceptions to the rules, but in this respect it is suggested that the reader consult the vast volume of more scientific papers or some of the thorough reviews.

Hydrocolloids are water-soluble polymers with an ability to thicken or gel aqueous systems. They can be classified according to origin, isolation method, function, texture, thermoreversibility, gelling time, or charge, but texture, thermoreversibility and gelling time are classification methods more appropriate for gelling agents.

Table 11.1 lists a number of hydrocolloids classified in these ways and it must be noted that the charge of proteins is dependent on the system pH.

The skeleton structures of hydrocolloids are shown in Figure 11.1. The characteristics of hydrocolloids with some examples are listed in Table 11.2, and Table 11.3 indicates the properties of food hydrocolloids in detail. The applications to which food hydrocolloids are put are listed in Table 11.4, and world consumption of human and pet food is outlined in Table 11.5.

Generally, hydrocollids show an increase in consumption of 2–3% p.a., however, low methoxyl pectin and xanthan gum show stronger growth, whereas the consumption of PGA is stagnant. LBG and gum arabic show a fluctuating trend, which in the case of gum arabic is caused by political instability.

As a retrospective example of how the information contained in this paper can be used, gelled milk desserts (pudding, flans, custard) are selected. Table 11.6 lists the various needs that have occurred over the time and the fulfilment of these.

Table 11.1 Classification of important hydrocolloids

Hydrocolloid	Origin				Isolation method			Function		
	Animal	Land-plants	Seaweed	Microorganism	Ground plant material	Exudate	Extract	Gelling agent	Thickener	Stabiliser
HM-pectin		x					x	x	x	x
LMC-pectin		x					x	x		
LMA-pectin		x					x	x		
Carrageenan			x				x	x		
Agar-agar			x				x	x	x	x
Alginate			x				x	x		
CMC		x							x	x
Guar gum		x			x				x	
LBG		x			x				x	
Xanthan gum				x				x	x	
Gellan gum				x				x	x	x
Gelatin	x						x	x		
Gum arabic		x				x				
Caseinate	x							x	x	x
Whey protein	x							x		
Soy protein		x					x	x		

HM-pectin = High methoxyl pectin
LMC-pectin = Low methoxyl conventional pectin
LMA-pectin = Low methoxyl amidated pectin
CMC = Carboxyl methyl cellulose
LBG = Locust bean gum

Table 11.1 *Continued*

Hydrocolloid	Texture			Thermo-reversibility		Gelling time		Charge		
	Brittle	Cohesive	Spreadable	Yes	No	Yes	No	−	0	+
HM-pectin	x	x			x	x		x		
LMC-pectin	x	x	x	x			x	x		
LMA-pectin	x	x	x	x			x	x		
Carrageenan	x	x	x	x			x	x		
Agar-agar	x			x			x		x	
Alginate	x		x		x		x	x		
CMC								x		
Guar gum									x	
LBG									x	
Xanthan gum		x		x			x		x	
Gellan gum	x			x			x		x	
Gelatin		x		x		x		x	x	x
Gum arabic									x	
Caseinate	x				x		x	x	x	x
Whey protein	x				x		x	x	x	x
Soy protein	x				x		x	x	x	x

HM-pectin = High methoxyl pectin
LMC-pectin = Low methoxyl conventional pectin
LMA-pectin = Low methoxyl amidated pectin
CMC = Carboxyl methyl cellulose
LBG = Locust bean gum

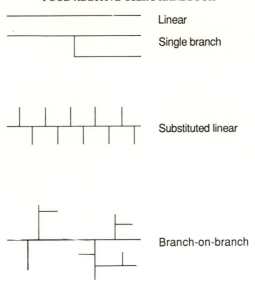

Figure 11.1   Structure of hydrocolloids.

Table 11.2   Characteristics and examples of hydrocolloids

Structure	Characteristics	Examples
Linear	Usually not more than two copolymerised sugar units. High viscosity. Unstable solutions. Difficult to dissolve. Risk of precipitation after dissolution (gelation)	Cellulose, amylose, pectin, carrageenan, alginate, agar
Single branch	Sugar units condensed with carbon groups other than C-1 or C-4	Dextrane
Substituted linear	Numerous short branches often consisting of only one sugar unit in length	Locust bean gum, guar gum
Branch-on-branch	Side chains on side chains. More stable and less viscous than linear. Typically, two or more types of sugar make up the polysaccharide. Excellent adhesive properties	Amylopectin, gum arabic

Property	High methoxyl pectin	Low methoxyl pectin	Kappa (κ) carrageenan	Iota (ι) carrageenan	Lambda (λ) carrageenan	Agar-agar	Alginate
Solubility in water	Sol. cold and hot	Sol. cold and hot	Sol. above 70°C Na⁺ and NH⁴⁺ Sol. cold	Sol. above 70°C Na⁺ and NH⁴⁺ Sol. cold K⁺ and Ca²⁺ swell cold to thixotropic dispersions	Sol. cold and hot	Sol. above 90°C	K⁺, Na⁺, NH⁴⁺ Sol. cold and hot Ca²⁺ insol. at neutral pH
Solubility in milk	Sol. cold and hot	Sol. hot	Sol. above 70°C	Sol. above 70°C	Sol. hot, swells cold	Sol. above 90°C	Insol. Na⁺ swells in boiling milk. Sol. with sequestering agents
Solubility in salt solutions	Insoluble	Insoluble	Insoluble	Sol. hot	Sol. hot	Sol. above 90°C	Insoluble
Solubility in sugar solutions	Sol. hot	Sol. hot	Sol. hot	Insoluble	Sol. hot	Sol. above 90°C	Sol. hot
Solubility in ethanol	Insol. above 20%	Insol. above 20%	Insol. above 20%	Insol. above 20%	Insol. above 20%	Insol. above 20%	Insol. above 40%
Other factors influencing solubility	Increases with decreasing MW, increasing randomness of COOH, decreasing sugar and Ca²⁺	Increases with decreasing sugar and Ca²⁺	Increases with decreasing Na⁺, K⁺ and Ca²	Increases with decreasing Na⁺, K⁺ and Ca²	Increases with decreasing Na⁺, K⁺ and Ca²		Increases with decreasing COOH, increasing pH, decreasing Ca²⁺
Solution viscosity	Low	Low	Low	Medium	High	Low	Low above pH = 5.5, high below pH = 5.5

Table 11.3 *Continued*

Property	High methoxyl pectin	Low methoxyl pectin	Kappa (κ) carrageenan	Iota (ι) carrageenan	Lambda (λ) carrageenan	Agar-agar	Alginate
Optimum pH range	2.5–4.0 $pK_a = 3.3$	2.5–5.5	4–10	4–10	4–10	2.5–10	2.8–10 $pK_a = 3.4$–4.4
Optimum soluble solids range	55–80%	30–80%	0–40%	0–20%	0–80%	0–80%	0–80%
Gelation conditions	pH below 4 and sol. solids 55–80%	Presence of Ca^{2+} 10–70 mg g^{-1} pectin. Temp. below setting temp.	Presence of K^+, Na^+ or Ca^{2+}. Temp. below setting temp.	Presence of K^+, Na^+ or Ca^{2+}. Temp. below setting temp.	Non-gelling	Temperature below 32–39°C	pH below 4 or presence of Ca^{2+} 20–70 mg per g alg.
Gel characteristics — Texture	Cohesive, no syneresis thermo-irreversible	Cohesive to brittle. Brittleness increases with increasing Ca^{2+} and decreasing sugar. Thermo-reversible	Strong, brittle. Brittleness increases with increasing K^+ and Ca^{2+} and decreasing LBG. Thermo-reversible	Soft, cohesive, thixotropic. Thermo-reversible. Thixotropy is lost with addition of LBG	Non-gelling	Strong, brittle. Thermo-reversible. Brittleness increases with increasing sugar	Acid gels soft, cohesive and thixotropic. Calcium gels strong, brittle. Thermo-irreversible

— Setting temp.	Increases with increasing DE, decreasing pH and increasing sugar	Increases with decreasing DE, increasing Ca²⁺, and increasing sugar	Increases with increasing K^+, Na^+, Ca^{2+} and sugar	Increases with increasing K^+, Na^+, Ca^{2+}, sugar and LBG	Non-gelling	Constant	Non-existent
— Gel strength	Increases with increasing concentration and MW	Increases with increasing concentration and Ca^{2+}	Increases with increasing concentration, K^+, Ca^{2+} and LBG	Increases with increasing concentration, K^+, Na^+ and Ca^{2+}	Non-gelling	Increases with increasing concentration, increasing sugar and increasing pH	Increases with increasing concentration Ca^{2+} and decreasing pH down to 3.6
Effect on milk at neutral pH	Precipitation	Gelation	Ionic interaction. Increased gel strength	Ionic interaction. Increased gel strength	Ionic interaction. Increasing viscosity	None	None. Insoluble
Effect on milk and other proteins at acid pH	Adsorption to case in particles below pH = 4.2. Adsorption to soy protein particles below pH = 4.8	None	Precipitation below iso-pH	Precipitation below iso-pH	Precipitation below iso-pH	None	None
Incompatibility	Water soluble alcohols, ketones, heavy metals, quaternary detergents, cationic macro-molecules	Water soluble alcohols, ketones	Water soluble alcohols, ketones, quaternary detergents, cationic macro-molecules	Water soluble alcohols, ketones, quaternary detergents, cationic, macro-molecules	Water soluble alcohols, ketones, quaternary detergents, cationic, macro-molecules	Water soluble alcohols, ketones	Water soluble alcohols, ketones, milk gum arabic

Table 11.3 *Continued*

Property	Propylene glycol alg.	CMC	Locust bean gum	Guar gum	Xanthan gum	Gelatine	Gum arabic
Solubility in water	Sol. cold and hot	Sol. cold and hot	Sol. above 85°C	Sol. cold and hot	Sol. cold and hot	Sol. above 40°C	Sol. cold and hot
Solubility in milk	Sol. cold and hot	Insoluble	Sol. above 85°C	Sol. cold and hot	Sol. cold and hot	Sol. above 40°C	Sol. cold and hot
Solubility in salt solutions	Insoluble	Insol. High DS types sol.	Sol. above 85°C	Sol. cold and hot	Sol. cold and hot	Sol. above 40°C	Sol. cold and hot
Solubility in sugar solutions	Sol. cold and hot	Sol. cold and hot	Sol. above 85°C	Sol. cold and hot	Sol. cold and hot	Sol. above 40°C	Sol. hot
Solubility in ethanol	Insol. above 40%	Insol. above 30%	Insol. above 20%	Insol. above 20%	Insol. above 50%	Insol. above 20%	Insol. above 60%
Other factors influencing solubility	Increases with decreasing MW, increasing pH, increasing divalent cations	Increases with decreasing sugar and Ca^{2+}				Increases with decreasing MW	Increases with increasing pH up to 6
Solution viscosity	High	High	High up to 85°C	High cold, low hot	High below 100°C	Low	Low
Optimum pH range	2.8–10	3–10 $pK_a = 4.2–4.4$	4–10	4–10	1–13	4.5–10 iso- pH = 4.8–5.2 (limed) iso- pH = 6.0–9.5 (acid)	2–10

Optimum soluble solids range	0–80%	0–80%	0–80%	0–80%	0–80%	0–80%	0–80%
Gelation conditions	Non-gelling	Non-gelling (gelation may occur with trivalent cations)	Non-gelling	Non-gelling	Presence of LBG, tara gum, cassia gum. Temp. below setting temp.	Temp. below setting temp.	Non-gelling
Gel characteristics — Texture	Non-gelling	Non-gelling	Non-gelling	Non-gelling	Cohesive, gummy thermo-reversible. Guar makes texture of xanthan/LBG gel more brittle	Soft to strong, cohesive, gummy. Thermo-reversible	Non-gelling
— Setting temp.	Non-gelling	Non-gelling	Non-gelling	Non-gelling	Constant	Increases with increasing MW and maturing temp.	Non-gelling
— Gel strength	Non-gelling	Non-gelling	Non-gelling	Non-gelling	Increases with increasing concentration	Increases with increasing concentration, and decreasing salt	Non-gelling

Table 11.3 *Continued*

Property	Propylene glycol alg.	CMC	Locust bean gum	Guar gum	Xanthan gum	Gelatine	Gum arabic
Effect on milk at neutral pH	None	Precipitation	Separation	Separation	None	None	None
Effect on milk and other proteins at acid pH	None	Adsorption to casein particles below pH = 4.6. Adsorption to soy protein particles below pH = 5.0	None	None	Precipitation below iso-pH	None	None
Incompatibility	Water soluble alcohols, ketones	Water soluble alcohols, ketones, quaternary detergents, cationic macro-molecules	Water soluble alcohols, ketones	Water soluble alcohols, ketones	Water soluble alcohols, ketones, gum arabic below pH = 5	Water soluble alcohols, ketones, anionic macro-molecules	Water soluble alcohols, ketones, alginate gelatine, xanthan gum

Table 11.4 Food applications of hydrocolloids

Hydrocolloid	Food	Function	Concentration (%)
Pectin	Jams, jellies, preserves	Gelation, thickening	0.1–1.0
	Bakery fillings, glazings	Gelation, thickening	0.5–1.5
	Fruit preparations	Thickening, stabilisation	0.1–1.0
	Fruit beverages, sauces	Thickening, stabilisation	0.01–0.5
	Confectionery	Gelation, thickening	0.5–2.5
	Dairy products	Stabilisation, gelation	0.1–1.0
Carrageenan	Ice cream	Stabilisation	0.01–0.03
	Chocolate milk	Stabilisation	0.01–0.03
	Flans and puddings	Gelation, thickening	0.1–0.5
	Liquid coffee whitener	Thickening	0.1–0.2
	Low-calorie jams	Gelation, thickening	0.8–1.2
	Dessert gels	Gelation	0.6–1.1
	Tart glazing	Gelation	0.8–1.0
	Meat glazing	Gelation, waterbinding	0.3–0.5
	Pimiento paste	Gelation	1.5–3.0
	Salad dressing	Stabilisation	0.3–0.6
Agar-agar	Icings	Gelation	0.1–0.3
	Confectionery	Gelation	0.3–1.8
	Meat products	Gelation	0.5–2.0
	Dairy products	Gelation	0.05–0.9
Alginate	Ice cream	Stabilisation	0.1–0.5
	Icings	Gelation	0.1–0.5
	Toppings	Gelation	0.3–0.5
	Salad dressings	Stabilisation	0.2–0.5
	Beer	Stabilisation	0.004–0.008
	Fruit drinks	Stabilisation	0.1–0.3
	Restructured foods	Gelation	0.6–1.0
	Simulated fruit	Gelation	0.8–1.0
CMC	Ice cream	Stabilisation, thickening	0.1–0.3
	Ripples	Thickening	0.1–0.4
	Sour milk	Stabilisation	0.1–0.2
	Cake mixes	Moisture retention	0.2–0.4
	Icings	Thickening, waterbinding	0.1–0.2
	Batters	Thickening, stabilisation	0.2–0.4
	Dry-mix beverages	Thickening	0.1–0.3
	Syrups	Thickening	0.2–0.6
LBG	Ice cream	Stabilisation	0.2–0.3
	Cream cheese	Thickening, moisture control	0.3–0.6
	Dessert gels	Gelation, water retention (together with carrageenan)	0.3–0.6
Guar gum	Ice cream	Stabilisation	0.2–0.3
	Cottage cheese	Thickening	0.3–0.6
	Processed cheese	Moisture retention	0.2–0.4
	Cake mixes	Thickening	0.1–0.2

Table 11.4 *Continued*

Hydrocolloid	Food	Function	Concentration (%)
Xanthan gum	Bakery jellies	Gelation, thickening	0.1–0.3
	Fruit drinks	Pulp suspension	0.02–0.06
	Cream cheese	Gelation	0.1–0.2
	Baked goods	Moisture retention	0.1–0.2
	Dressings	Stabilisation	0.2–0.3
Gelatine	Yoghurt	Gelation	0.3–1.0
	Dessert gels	Gelation	4–6
	Confectionery	Gelation	3–10
	Meat products	Gelation	1–5
	Mousses	Stabilisation	1–3
	Minarine	Stabilisation	1–3
Gum arabic	Flavour fixation	Encapsulation	80–90
	Confectionery	Stabilisation	10–60
	Flavour emulsions	Stabilisation, emulsification	10–30

Table 11.5 World consumption of hydrocolloids in human and pet food

Hydrocolloid	Consumption 1986 (tonnes)	Trend
High methoxyl pectin	12 000	Increasing 2–3% p.a.
Low methoxyl pectin	5 000	Increasing 7–8% p.a.
Carrageenan	14 000	Increasing 2–3% p.a.
Agar-agar	6 000	Increasing 2–3% p.a.
Alginate	6 000	Increasing 2–3% p.a.
Propylene glycol alg.	2 000	Increasing 0.1% p.a.
CMC	7 000	Increasing 2–3% p.a.
LBG	6 000	Fluctuating
Guar gum	16 000	Increasing 2–3% p.a.
Xanthan gum	4 000	Increasing 5–6% p.a.
Gelatin	40 000	Increasing 1–2% p.a.
Gum arabic	20 000	Fluctuating

Table 11.6 The use of hydrocolloids to solve problems in the food industry

Needs	Solution	Problems
After dinner dessert made at home	Milk pudding gelled with amylose	Retrogradation (linear); needs preparation; heavy
Pudding with no retrogradation (cracking) made at home	Introduction of waxy starches and modified starches (branched)	Needs preparation; heavy
Lighter pudding with no retrogradation made at home	Substitution of part of or all starch with carrageenan, agar-agar, gelatin. (soluble in milk, gelling agents, low use concentration)	Needs preparation

Table 11.6 *Continued*

Needs	Solution	Problems
More convenience	Instant pudding with carrageenan, alginate, xanthan gum	Taste of polyphosphates, texture not as gelled as the original; still some preparation
Convenience, better taste and improved texture	Industrially produced, pasteurised puddings with carrageenan, gelatine. (Soluble at pasteurising temperature, gelling agents)	Shelf-life
Convenience, improved shelf life	Sterilised puddings with carrageenan, agar-agar, gelatin. (soluble at sterilising temperature, gelling agents)	Increased competition
Varieties	Multi-layer desserts with carrageenan, gelatine, agar-agar	Long processing time; each layer must be cooled before addition of next layer; gelatine further requires time for gelation
Varieties with shorter production time	Cold filling using carrageenan (thixotropy)	Boundary problems when combining acid water gel with neutral milk gel. (Carrageenan precipitates milk protein)
Varieties with no boundary problem	Substituting carrageenan in milk gel with low methoxyl pectin (setting via slow release of Ca^{2+})	Increased competition
Varieties	Sour milk desserts with agar-agar, gelatine, low methoxyl pectin	

Further reading

Copenhagen Pectin, Carrageenan general description, Booklet B1.
Copenhagen Pectin, Pectin general description, Booklet B4.
Dickinson, E. and Stainsby, G. (1982). *Colloids in Food*. Applied Science Publishers, London and New York.
Glicksman, M, (1982, 1983, 1986). *Food Hydrocolloids*, Vols 1, 2 and 3. CRC Press, Inc., Boca Raton, Florida.
Graham, H. D. (1977). *Food Colloids*. The AVI Publishing Company INC. Westport, Connecticut.
Lawrence, A. A. (1976). *Natural Gums for Edible Purposes*. Noyes Data Corporation. Park Ridge, New Jersey, USA.
Whistler, R. L. (1973). *Industrial Gums*. Academic Press. New York, San Francisco, London.

12 Antifoams and release agents

A. A. ZOTTO

12.1 Antifoams

Foam is a mass of bubbles created when certain types of gas are dispersed into a liquid and the dispersion is then stabilised.

While the actual cause of foam is a complicated study in physical chemistry, its existence presents serious problems in both the operation of industrial processes and the quality of finished products. If not properly controlled, foam can reduce equipment capacity and increase processing time and expense. Foam can be controlled by making basic changes in the process itself or by using mechanical defoaming equipment. However, chemical defoamers have proved to be the most effective and economical.

An effective chemical defoaming agent must:

- possess lower surface tension than the system to which it is added;
- disperse readily in the system;
- possess poor or low solubility (incompatibility) in the system;
- be inert;
- leave no substantial residue or odour;
- meet regulatory requirements where applicable.

These requirements are met most effectively with silicone antifoams. Silicone antifoams are extensively used where product or process foaming problems are encountered, and some of the factors to be considered in the selection process are:

- chemical nature of the foam forming agent;
- foaming tendency of the agent;
- solubility and concentration;
- presence of electrolytes, colloids or other surface active agents;
- temperature, pH and viscosity of the system;
- processing equipment involved;
- end-use of product containing the antifoam.

12.1.1 Available products

Antifoams are available in three basic forms for matching specific products to specific performance requirements (Table 12.1).

Table 12.1 Forms of antifoam

Non-aqueous		Aqueous
Fluids	Compounds	Emulsions
SF18-350*	AF70*	AF72*
	AF9000*	AF75*
		AF9020*

*Comply with FDA Regulations for use in food processing applications.
Data relates to GE Silicone products.

Aqueous antifoams: Emulsions. These are water-based products which provide easy dispersibility for maximum defoaming efficiency.

Non-aqueous antifoams: Fluids and compounds. These are fluid products and formulations of fluids containing specially prepared fillers.

12.1.2 Product descriptions

Silicone fluid (SF18-350) (Table 12.2). This fluid is a polydimethylsiloxane polymer with a kinematic viscosity of $3.5 \times 10^{-4} m^2 s^{-1}$. The structure consists of the siloxane linkage backbone with pendant and terminal methyl groups.

$$CH_3-Si(CH_3)_2-[O-Si(CH_3)_2]_x-O-Si(CH_3)_2-CH_3$$

The molecular weight of this polymer is approximately 15 000.

Table 12.2 Typical product data (fluids)

Property	Value SF18-350
Silicone content (%)	100
Density (kg m^{-3})	9.58×10^2
Specific gravity at 25°C/25°C	0.97
Kinematic viscosity (m^2 s^{-1})	3.5×10^{-4}
Flash point (°C) (Pensky-Martens closed cup)	204
Surface tension (N m^{-1} at 25°C)	21.1×10^{-3}

Silicone compounds (AF70, AF9000) (Table 12.3). An antifoam compound is composed of a small amount of high surface area fused silica filler dispersed through and suspended in the SF18-350 silicone fluid. The addition of the filler provides a pseudo-mechanical method of providing a further reduction of the surface tension of the foam formation. In effect, the filler acts as a flaw

Table 12.3 Typical product data (compounds)

Property	Value	
	AF70	AF9000
Silicone content (%)	100	100
Density (kg m^{-3})	1.0×10^3	1.0×10^3
Specific gravity at 25°C/25°C	1.01	1.01
Dynamic viscosity, 25°C Pa.s max.	1.5	1.5
Flash point (°C Pa.s) open cup	315	315

in the surface of the bubble, thereby enabling an easier rupture of the bubble.

Silicone emulsions (AF72, AF75, AF9020) (Table 12.4). Emulsification of SF18-350 fluid at different solids concentrations and with different emulsifier systems provides these products. AF9020 emulsion works over the broadest range of pH and ionic/non-ionic systems. For any specific situation however, either AF72 or AF75 emulsions might offer a performance advantage.

Table 12.4 Typical product data (silicone antifoam emulsions)

Property	Values		
	AF72	AF75	AF9020
Total solids	44.2 ± 1	13.75 ± 1.25	28.75 ± 1.25
Silicone content (%)	30	10	20
Density (kg m^{-3})	1.0×10^3	1.0×10^3	1.0×10^3
Specific gravity at 25°C/25°C	1.01	1.02	1.01
Dynamic viscosity at 25°C, Pa.s max.	1.5	3.0	3.5
Colour	White		
Heat stability	Stable to 43°C		
Dispersibility	Readily dispersible in cold water with mild agitation		
Dilution stability	Less than 2% creaming and no settling after 24 h at 10% silicone content		
Emulsifier type	Non-ionic		

12.1.3 Processing guidelines

When evaluating antiform emulsions or fluids, it is suggested that the evaluation be started at about 10 ppm silicone. This starting point represents an average level of silicone found to be effective in many applications. The level can then be adjusted in either direction to determine the minimum effective concentration for the specific system requiring foam control.

For applications related to food processing, consult applicable FDA or USDA regulations for the maximum permissible level of silicone allowed.

Table 12.5 Products for food applications

Food processing	SF18-350	AF70	AF9000	AF72	AF75	AF9020
Fermentation	X	X		X	X	X
Antifoam formulating food grade	X	X	X	X	X	X
Brine systems				X	X	X
Cheese whey processing				X	X	X
Chewing gum base				X	X	X
Corn oil manufacture	X	X	X			
Deep-fat frying	X	X	X			
Esterification of vegetable oil	X	X	X			
Fermentation systems				X	X	X
Fruit processing				X	X	X
Instant coffee and tea manufacture				X	X	X
Jam and jelly making				X	X	X
Juice processing				X	X	X
Pickle processing				X	X	X
Potato processing				X	X	X
Rice processing				X	X	X
Sauce processing				X	X	X
Soft drink processing				X	X	X
Sugar refining				X	X	X
Syrup manufacture				X	X	X
Vegetable processing				X	X	X
Wine making				X	X	X
Yeast processing				X	X	X

Data relate to GE Silicone products

Table 12.6 Equivalent measures for defoamers

Parts per million	Percent	Ounces per 1000 gallons	Ounces per 1000 pounds	Grams per 1000 litres	Grams per 1000 kilograms
1	0.0001	0.134	0.016	1.09	1.09
10	0.0010	1.340	0.160	10.90	10.90
100	0.0100	13.400	1.600	109.00	109.00
1000	0.1000	134.000	16.000	1090.00	1090.00

Data relate to GE Silicone products

12.2 Release agents

Silicone emulsion (Table 12.8), e.g. SM2128, based on silicone fluid (SF18-350), is a typical release emulsion useful with materials coming in incidental contact with food. Its key performance properties are as follows: • easy application, • good lubricity, • easy release, • high temperature stability, • low volatility • chemically inert • low concentrations • water dispersible.

Table 12.7 Areas in which grades of products may be used as direct and/or indirect food additives

		AF 70	AF 72	AF 75	AF 9000	AF 9020	SF 18-350
Part 145	**Canned Fruits**						
145.180	Canned pineapple		x	x	x		
Part 146	**Canned Fruit Juices**						
146.185	Canned pineapple juice		x	x	x		
Part 150	**Fruit Butters, Jellies, Preserves, etc.**						
150.?	Fruit butters, jellies, etc.		x	x	x		
Part 173	**Secondary Direct Food Additives, Human**						
173.340	Defoaming agents	x	x	x	x	x	x
Part 175	**Indirect Food Additives**						
175.105	Adhesives	x	x	x	x	x	x
175.300	Resinous and polymeric coatings	x			x		x
Part 176	**Indirect Food Additives; Paper and Paperboard Components**						
176.170	Components in contact with aqueous and fatty foods	x	x	x	x	x	x
176.180	Components in contact with dry food	x	x	x	x	x	x
176.200	Defoaming agents used in coatings	x	x	x	x	x	x
176.210	Defoaming agents – paper and paperboard	x	x	x	x	x	x
Part 177	**Substances for Single or Repeated Use Food Contact Surfaces**						
177.1200	Cellophane	x	x	x	x	x	x
177.2260	Fillers; resin bonded	x	x	x	x	x	x
177.2800	Textiles and textile fabrics	x	x	x	x	x	x
Part 178	**Indirect Food Additives; Adjuvants; Production Aids, etc.**						
178.3120	Animal glue	x	x	x	x	x	x
178.3570	Lubricants with incidental food contact	x		x	x		x
Part 181	**Prior-sanctioned Food Ingredients**						
181.28	Release agents						x
Part 701	**Cosmetic Labelling**						
701.3	Designation of ingredients	x	x	x	x	x	x

Note: Part numbers refer to the Code of Federal Regulations, Title 21, Chapter 1.
Data relate to GE Silicone products.

12.2.1 Processing guidelines

SM2128 emulsion can be applied by any conventional means such as swabbing, brushing or spraying. In most applications where SM2128 emulsion is used as a mould release agent, it should be diluted with water

Table 12.8 Typical product data

Emulsion	SM2128
Emulsifier type	Non-ionic
Base oil	SF18-350
Silicone content (%)	35
Viscosity of contained oil at 25°C (m^2 s^{-1})	3.5×10^{-4}
Density (kg m^{-3})	9.82×10^2
Storage stability (months)	12
Diluent	Water
Total solids (%)	37–41
Colour	White

Data relate to GE Silicone products.

before use. The actual concentration of silicone required will vary with the specific moulded part to be released. However, in most cases, concentrations of 0.5–2.0% silicone emulsion SM2128 with water have been found most effective as a release agent for paper products such as paper plates and cups. It is also effective as a lubricant and release agent for rubber parts such as gaskets, grommets, bushings and plastisol compounds.

The following list summarises the areas in which SM2128 emulsion may be used as an indirect food additive:

Part 176   Indirect Food Additives; Paper and Paperboard Components
176.170   Components in contact with aqueous and fatty foods
176.180   Components in contact with dry food

Part 177   Substances for Single or Repeated Use Food Contact Surfaces
177.1200 Cellophane

Part 178   Indirect Food Additives; Adjuvants; Production Aids, etc.
178.3120 Animal glue

Further reading

Anon (1986) Foam control using foaming and/or defoaming agents. *Food Processing* **47** (12), 42–44.

Anon (1988) Baking: Lipids were evaluated as release agents in the food industry. *Alimenta*, **27** (1), 11.

Anon (1988) Refined lecithin pan release agents are easy to use and multifunctional. *Food Processing* **49** (5), 78–80.

Combs, C. (1990) Choosing the right antifoam agent for the beverage line. *Beverage Industry* **81** (5), 11.

Lewis, R. J. (1989) *Food Additives Handbook*, Van Nostrand Reinhold, New York.

13 Flour improvers and raising agents

D. BIEBAUT

13.1 Dough conditioners in yeast-raised systems

A good dough conditioner controls every part of the baking process (Table 13.1) and influences all the important parameters of the bread, such as: crumb structure; softness of the crumb; crustiness; volume; and shelf-life.

The active ingredients may consist of: oxidising agents (Table 13.2); reducing agents (Table 13.2); emulsifiers (and triglycerides) (Tables 13.4, 13.5 and 13.6); sugars (Table 13.7); enzymes (Tables 13.8 and 13.9); inorganic salts (Table 13.10), flour, other than wheat; and gluten (Table 13.11).

All these ingredients have an influence on: the formation of the gluten network; the production of CO_2; and the retention of CO_2.

Table 13.1 Additives in baking

Additive	Function
Acesulfame-K	Artificial sweetener[1]
Agar-agar	Stabiliser, thickener
Ammonium bicarbonate	Leavening
Ammonium chloride	Yeast nutrient
Ammonium sulfate	Yeast nutrient
Amylase, fungal enzyme	Converts starch to sugar, aids fermentation
Ascorbic acid	Oxidant, aids bread processing
Aspartame	Artificial sweetener[2]
BHA	Antioxidant, retards rancidity
BHT	Antioxidant, retards rancidity
Calcium carbonate	Calcium source, buffering
Calcium caseinate	Milk derivative
Calcium peroxide	Oxidant, aids bread processing
Calcium propionate	Preservative, retards mould growth
Calcium stearoyl lactylate	Dough strengthener, better texture
Calcium sulfate	Source of calcium
Caramel	Colouring
Carob bean	Stabiliser, thickener
Carrageenan	Stabiliser, thickener
Cellulose gum	Stabiliser, thickener
Citric acid	Antioxidant, retards rancidity
Cream of tartar	Leavening
Diacetyl tartaric acid esters of mono- and diglycerides (DATEM)	Emulsifier

Table 13.1 *Continued*

Additive	Function
Ethoxylated mono- and diglycerides	Dough strengthener, better texture
Gelatin	Stabiliser, thickener
Iron	Mineral nutrient
L-cysteine	Reducing agent for bread
Lecithin	Emulsifier
Malt, malted barley flour	Enzyme, aids processing
Mono- and diglycerides	Emulsifier and crumb softener, retards staling
Monocalcium phosphate	Leavening, dough conditioning
Niacin	Important vitamin
Pectin	Thickener, stabiliser
Polydextrose	Low-calorie carbohydrate, bulking agent
Polyglycerol esters	Emulsifier and crumb softener, retards staling
Polysorbate 60, 65	Emulsifier, aids whipping, retards staling
Potassium bromate	Oxidant, aids bread processing
Potassium sorbate	Preservative, retards mold growth
Propylene glycol monoesters	Emulsifier and crumb softener, retards staling
Propyl gallate	Antioxidant, retards rancidity
Protease, fungal enzyme	Enzyme, softens protein, aids bread processing
Riboflavin	Important vitamin
Sodium acid pyrophosphate	Leavening
Sodium alginate	Stabiliser, thickener
Sodium benzoate	Preservative, retards mould growth
Sodium bicarbonate (baking soda)	Leavening
Sodium bisulphite, sodium metabisulphite	Dough conditioner, aids processing
Sodium caseinate	Milk derivative
Sodium diacetate	Preservative, retards mould growth
Sodium propionate	Preservative, retards mould growth
Sodium stearoyl lactylate	Dough strengthener, better texture
Sorbic acid	Preservative, retards mould growth
Sorbitan monostearate	Emulsifier and crumb softener, retards staling
Starch, corn starch	Thickener
Sucrose polyesters	Emulsifier
Succinylated monoglycerides	Dough strengthener, better texture
TBHQ	Antioxidant, retards rancidity
Thiamin mononitrate	Important vitamin
Vinegar	Acidifying and mould retarding
Wheat gluten	Wheat protein concentrate

[1] Use in dry bases for gelatins, puddings and bases approved by Food and Drug Administration (FDA) July 1988.
[2] New uses in beverages and gelatin desserts approved by FDA in June 1988; bakery applications still pending.

The gluten network that is formed during mixing is responsible for both strength and extensibility. The ingredients of a bread improver that have an effect on these properties are, especially, oxidants and reducing agents. By adding oxidants we increase the strength of the dough. Table 13.2 shows the difference in reaction rate between oxidizing agents.

Table 13.2 Oxidising and reducing agents

OXIDISING AGENTS

Compound	Rate of reaction	Point of addition	Conditions for reaction	FDA limits (ppm)
Calcium peroxide	Fast	Mixer	Wet	75
Potassium iodate	Fast	Mixer	Wet	75
Azodicarbonamide (ADA)	Fast	Mixer	Wet	45
Ascorbic acid	Medium-fast	Mixer	Wet – flour contact	GMP[1]
Calcium iodate	Intermediate	Make-up	Wet	75
Coated ascorbic acid	Intermediate	Floor time	Wet – flour contact	GMP
Potassium bromate	Slow	Late proof and oven	Acidification, heat	75
Calcium bromate	Slow	Late proof and oven	Acidification, heat	75

REDUCING AGENTS

Compound	Rate of reaction	Point of addition	Conditions for reaction	FDA limits (ppm)
L-cysteine hydrochloride	Fast	Mixer	Wet	GMP
Sodium metabisulphite	Fast	Mixer	Wet	GMP
Sorbic acid	Fast	Mixer	Wet	GMP
Coated ascorbic acid	Fast	Mixer	Wet – closed system	GMP
Fungal protease	Slow	Sponge mixer	Wet – low salt time	GMP

[1] GMP = Good Manufacturing Practice.

Table 13.3 Effect of oxidation on dough and bread

UNDER-OXIDISED	OVER-OXIDISED
Dough	*Dough*
Weak	Tight
Soft	Firm
Sticky	Bucky
Extensible	Difficult to mould
Difficult to machine	Tears easily
Bread	*Bread*
Poor volume	Small volume
Weak crust	Rough break-and-shred
Uneven grain and texture	Rough crust
Poor break-and-shred	Uneven grain
Poor symmetry	Large holes

Besides speed, there are other differences: for instance, ADA is very sensitive to overmixing, $KBrO_3$ is sensitive to overdosage, and although ascorbic acid is less sensitive to overdosage, it is considered to be less powerful than potassium bromate.

Reducing agents are mostly used in combination with oxidants (Table 13.2). They act as dough softeners and improve the extensibility, making doughs of certain strong flours more machinable. They are also able to reduce the mixing time needed for a good dough formation.

In bread-making processes where the dough is frozen and stored for some time, the liberation of glutathion by damaged yeast cells during freezing and thawing may require the use of vital wheat gluten. Vital wheat gluten is also used for the production of whole wheat bread, where the bran fraction tends to make the formation of the protein network more difficult.

Emulsifiers are known to have an influence on both proteins and the starch fraction. They are active in almost every part of the baking process. They especially provide security to the baking process and compensate for under- or overmixing, as long as this is not too excessive. Where the fermentation time is excessive, emulsifiers prohibit the dough from collapsing. This is of particular interest in processes where the dough is roughly treated by manual action or because of automation.

The emulsifiers that are permitted for use in bread and speciality breads are shown in Tables 13.4, 13.5 and 13.6. The legislation for the use of emulsifiers in bakery products varies from country to country, and the local laws should be checked for details.

- Distilled monoglycerides have unique properties in aqueous systems forming lamellar dispersions. They find wide use as starch complexing agents in bread and as aerating agents in fat-free cakes.
- Lecithins are mainly used in countries where no other emulsifiers are allowed. Although lecithins are weak emulsifiers, they can have a positive influence on weak flour.
- Monoglycerides, esterified with organic acids, are the most powerful emulsifiers used in bread improvers. The most important monoglycerides are:

(1) Datem (monoglycerides + acetic acid + tartaric acid)
   Advantages:
   highest bread volume
   increased crustiness
   gives a very good stability to the dough
(2) SSL and CSL (fatty acids + lactic acid, neutralised)
   Advantages:
   good volume
   good shelf-life
   enhanced crumb softness

Table 13.4   Application of emulsifiers in the European Economic countries (USA covered in Table 13.6)

Name	Belgium		Germany		Denmark	
	Dose (ppm)	Prod.	Dose (ppm)	Prod.	Dose (ppm)	Prod.
Mono and diglycerides of fatty acids	10 000	A/B/C	GMP	A/C D	GMP	A/B/ C/D
	GMP 10 000	2 M				
Mono- and diglycerides esterified with organic acids	GMP	2	20 000	A/C	GMP	A/B/ C/D 1
	10 000	A/B/C				
	10 000	M				
Sodium stearoyl-2-lactylate	5 000	B/C 3	–	–	3 000	A/B/ C/D
Calcium stearoyl-2-lactylate	5 000	B/C 3	–	–	3 000	A/B/ C/D
Lecithin			GMP	A/C		
	GMP	2		D	GMP	A/B/ C/D
	2 000	A/B				
	GMP	C				
	2 000	M				

Key: A = Ordinary bakery products
      B = Fine bakery products
      C = Pastry – biscuits
      D = Confectionery
      M = Margarines

To produce CO_2 during the fermentation, yeast needs fermentable carbohydrates. Besides adding sugars, enzymes may be added to produce fermentable sugars, during the bread-making process, from the starch fraction of flour.

Flour contains both α- and β-amylases. Although commercial flour contains sufficient β-amylase, it may require more α-amylase. There are various methods of adding α-amylase (Tables 13.8 and 13.9):

– malt flour
– fungal amylases (*Aspergillus*)
– bacterial amylases (*Bacillus* species)

Table 13.4 *Continued*

Spain		France		Italy		Netherlands		UK	
Dose (ppm)	Prod.	Dose (ppm)	Prod.	Dose (ppm)	Prod.	Dose (ppm)	Prod.	Dose (ppm)	Prod.
30 000	C	20 000	A[6]			GMP	C	GMP	
3 000	B/A	10 000	C	30 000	C/B	10 000	A[3]		
GMP	D					GMP	M		
30 000	C	10 000	C	2 000	A(6)	10 000	A[3]	GMP	[1]
			A[6]	30 000	C/B				
3 000	A/B								
10 000	M								
400	D								
5 000	C/B	5 000	C[4]	–	–	GMP	C	5 000	A
10 000	M/D								
5 000	C/B	5 000	A[5]	–	–	GMP	C	5 000	A
10 000	M					0.5%	A[3]		
1 000	D								
30 000	C			2 000	BCD	GMP	C	GMP	
2 000	A	3 000	A/B			5 000	A/B		
4 000	B	2 000	C			GMP	M		
GMP	D/M								

[1] Except E472d and f (prohibited) and E472e (limited to 5000 ppm in A/B/C/D).
[2] Raw materials intended for A/B/C as in tin-greasing agent.
[3] Level calculated on the basis of the flour.
[4] Industrial pre-wrapped products made with egg pastry.
[5] Special pre-wrapped bread.
[6] Special bread only.

Bacterial amylases can cause problems because they continue their activity after the bread has left the oven. Their level must therefore be carefully controlled, otherwise the bread crumb breaks down. To obtain optimal results, it is useful to add sugars and enzymes to the flour. In that way, the sugar level is assured, not only during the beginning of the fermentation, but also throughout the whole bread-making process.

Not only sugar-producing enzymes are useful in a bread improver. Soyflour contains lipoxygenase, acting as a bleaching agent on flour components.

Table 13.5　Dough conditioners and crumb softeners

Conditioner	Typical use levels[1] (%)	Estimated use[2]	
		USA	Western world
Calcium and sodium stearoyl lactylate (CSL, SSL)	0.25–0.5	8.0–12	12–17
Diacetyl tartaric acid esters of mono- and diglycerides (DATA)	0.25–0.5	0.5–0.7	No estimate
Ethoxylated mono- and diglycerides (EMG)	0.15–0.4	1–4	No estimate
Polysorbate 60 (PS-60)	0.15–0.4	0.5–0.9	0.5–1.5
Succinylated monoglycerides (SMG)	0.15–0.4	0.1–0.3	0.2–0.5
Distilled monoglycerides (DMG):		5–7	10–12
hydrated (25%)	1–1.5		
hydrated (50%)	0.5–0.75		
powdered (100%)	0.25–0.5		
Mono- and diglycerides (MD)	0.4–0.75	9–11	No estimate

[1] Percent of formula wheat flour.
[2] $\times 1000\,t\,yr^{-1}$.

Table 13.6　Baking additives with maximum level in the USA

Additive	Level limit
Ascorbic acid	GMP[1]
Azodicarbonanide	45 ppm[2]
Calcium diacetate	GMP
Calcium stearoyl lactylate	0.5%[2]
Cysteine	GMP
Diacetyl tartaric acid esters of mono- and diglycerides (DATEM)	GMP
Ethoxylated mono- and diglycerides	0.5%[2]
Ethylene/propylene oxide copolymer	0.5%[2]
$KBrO_3$, $CaBrO_2$, Ca IO, calcium peroxide	0.0075%[2]
Mono- and diglycerides	GMP
Mono-calcium phosphate (yeast food)	0.25%[2]
Polysorbate 60 in cakes	0.46% mix weight
Polysorbate 60 in bread	0.5%[2]
Polysorbate 65 in cakes	0.32% mix weight
Sodium diacetate	GMP
Sodium propionate	GMP
Sodium stearoyl fumarate	0.5%[2]
Sodium stearoyl lactylate	0.5%[2]
Sorbitan monostearate in cakes	0.61% mix weight
Sorbitan monostearate in icing	0.70% batch
Succinylated monoglycerides	0.5%[2]

[1] Good Manufacturing Practice.
[2] Based on flour.

Table 13.7 Sweeteners for baking

Sweetener	100 kg contain (kg):		Kilograms needed to get 100 kg of solids	100 kg of solids contain (kg):					Storage temperature (°F)	Fermentable solids (%)	Relative sweetness (dry basis)
	Solids	Water		Suc.	Fru.	Dex.	Mal.	Other			
Dry sucrose	100	—	100	100	—	—	—	—	—	100	1.0
Liquid sucrose	67	33	150	100	—	—	—	—	68+	100	1.0
Medium invert	77	23	130	46	25	27	—	2	68+	100	1–1.2
Total invert	77	23	130	3	45	48	—	4	90–95	100	1–1.2
Dry dextrose	91	9	110	—	—	100	—	—	—	100	0.75
Liquid dextrose	71	29	141	—	—	100	—	—	130+	100	0.75
95 DE corn syrup	71	29	141	—	—	94	3	3	130+	97	0.75
69 DE corn syrup	82	18	122	—	—	43	32	25	115	75	0.65
62 DE corn syrup	82	18	122	—	—	36	30	34	115	66	0.60
42 DE corn syrup	80	20	125	—	—	19	14	67	110	33	0.45
42% HFCS	71	29	141	—	42	52	2	4	90–95	96	1.0
55% HFCS	77	23	130	—	55	42	1	2	80–85	98	1–1.2
80% HFCS	77	23	130	—	80	19	—	1	68+	99	1.2–1.6

DE = Dextrose Equivalent. HFCS = High Fructose Corn Syrup.
Suc. = Sucrose; Fru. = Fructose; Dex. = Dextrose; Mal. = Maltose.

Table 13.8   Enzymes for baking

(a) α-Amylase preparations

Source	Form	Potency (α-amylase units)	Point of use
Barley malt	Flour	ca. $50\,g^{-1}$	Mill and bakery
Wheat malt	Flour	ca. $50\,g^{-1}$	Mill and bakery
Malt syrup	Viscous liquid	ca. $5-30\,g^{-1}$	Added to doughs
Dried malt extract	Powder	ca. $5-30\,g^{-1}$	Added to doughs
Fungal	Tablet	5000 per tablet	Added to doughs
Fungal	Powder	$50-200\,g^{-1}$	Added to doughs

Levels used in baking are typically 13 000 to 26 000 α-amylase units per 100 kg of flour

(b) Thermostability of α-amylase from various sources

Temperature		Percentage of enzyme activity		
°C	°F	Fungal	Wheat malt	Bacterial
65	149	100	100	100
70	158	52	100	100
75	167	3	58	100
80	176	1	25	92
85	185	—	1	58
90	194	—	—	22
95	203	—	—	8

(c) Major enzymes in fermentation

Enzyme	Source	Acts on	Products
α-Amylase (diastatic enzyme)	Flour Fungal enzyme Malt Bacterial enzyme	Starch	Soluble starch and dextrins
β-amylase	Flour Malt	Dextrins	Maltose
Invertase	Yeast	Sucrose	Invert sugar (dextrose and fructose)
Maltase	Yeast	Maltose	Dextrose
Zymase	Yeast	Invert sugar and dextrose	Carbon dioxide gas, alcohol and flavours
Proteinase (proteolytic enzyme or protease)	Flour Fungal enzyme Bacterial enzyme	Proteins (gluten)	Enables faster mixing and improved dough extensibility

Table 13.9   Commercial α-amylases

(a) Properties of amylases for commercial applications

Source	Amylase type	pH ranges		Temperature ranges	
		Optimum	Stability	Optimum	Effective
A. oryzae	α-Amylase	4.8–5.8	5.5–8.5	45–55°C(113–130°F)	Up to 60°C(140°F)
B. subtilis	α-Amylase	5.0–7.0	4.8–8.5	60–70°C(140–158°F)	Up to 90°C(195°F)
Barley malt	α-Amylase	4.0–5.8	4.9–9.1	50–65°C(122–150°F)	Up to 70°C(158°F)
Porcine pancreas	α-Amylase	6.0–7.0	7.0–8.8	45–55°C(113–130°F)	Up to 55°C(130°F)
Barley malt	β-Amylase	5.0–5.5	4.5–8.0	40–50°C(104–112°F)	Up to 55°C(130°F)
A. niger	Gluco-amylase	4.0–4.5	3.5–5.0	55–60°C(130–140°F)	Up to 70°C(158°F)

(b) Amylase baking results

	Malted wheat flour	Fungal enzyme	Bacterial enzyme
Flour adjusted to same gassing power			
Amount $(\text{mg}(100\,\text{g})^{-1}$ flour)	250.0	6.3	63.2
Loaf volume (ml)	3075	3188	2968
Grain (%)	96	90-open	70-open
Texture (%)	95	90	70
External appearance	Very good	Good	Very good
Remarks	Excellent dough	Soft dough	Sticky crumb
Flour adjusted to same maximum viscosity			
Amount $(\text{mg}(100\,\text{g})^{-1})$	250.0	224.0	29.5
Loaf volume (ml)	3075	—	3075
Grain (%)	95	—	80
Texture (%)	95	—	75
External appearance	Very good	—	Very good
Remarks	Excellent dough	—	Sticky crumb

[1] Dash = Sponge dough liquified, impossible to handle.

13.2 Chemical leavening

Chemical leavening depends upon the production of CO_2 due to the reaction between an acid and soda (sodium bicarbonate) when both are dissolved in water. This chemical reaction can be written as:

$$NaHCO_3 + HX \rightarrow NaX + CO_2 + H_2O$$

In reality, the leavening occurs in two steps. During mixing, air bubbles are incorporated into the batter. This aeration is followed quickly by the formation of CO_2 giving rise to a fine dispersion of small air bubbles and CO_2. The greater the dispersion and retention of the gas bubbles in the batter during baking, the finer the grain will be and the thinner will be the cell walls in the baked cake. Suitable emulsifiers in the formula will stabilise the product.

Since sodium bicarbonate will dissolve almost immediately in water to react with the available acid, the rate of dissolution of the acid will govern the rate of release of CO_2 from the bicarbonate. CO_2 must be released at just the right time during batter or dough preparation and during subsequent baking. The reaction of neutralising soda with an acid can be measured, and has been defined by the term 'neutralising value' (NV). It has been defined as the part by weight of sodium bicarbonate which will neutralize 100 parts by weight of an acid leavener to a specific end point, i.e. to convert all of the bicarbonate to CO_2 (Table 13.12)

$$NV = \frac{g \text{ sodium bicarbonate}}{g \text{ leavening acid}} \times 100$$

Table 13.10   Inorganic salts in relation to baking

(a) Salt: technological functions

Function	Action
Controls rate of fermentation	Salt acts as dough system buffer. Lessens yeast activity. Prevents development of bacterial action of 'wild' types of fermentation
Stabilises dough structure	Strengthens gluten. Helps ensure good dough handling properties on high-speed equipment

(b) Adjustments for water

Water type	Subclass	Yeast food type	Amount	Other treatment
Acid below 6.8 pH	Soft, over 120 ppm	Regular bromate	More than normal	Salt in sponge; Ca; soy in extreme cases
Acid below 6.8 pH	Hard, under 180 ppm	Regular bromate	Normal	None
Acid below 6.8 pH	Very hard, 180 to 1200 ppm	Regular bromate	Less than normal	Malt in sponge (in extreme cases)
Average pH 7.0 to 8.0	Soft	Regular	More than normal	None
Average pH 7.0 to 8.0	Moderately hard	Regular	Normal	None
Average pH 7.0 to 8.0	Hard	Regular	Less than normal	Malt in sponge

Table 13.10(b) *Continued*

Water type	Subclass	Yeast food type	Amount	Other treatment
Alkaline pH above 8	Soft	Regular, plus $CaHPO_4$ or acid	More than normal	$CaHPO_4$
Alkaline pH above 8	Moderately hard	Acid	Normal	None
Alkaline pH above 8	Hard	Acid	Less than normal	Much malt, plus acetic or lactic acid in extreme cases

Table 13.11 Wheat gluten applications

Baked product	Optimum level	Gluten function
Wheat with bran[1]	3.0% bakers	Improves uniformity of loaf volume, grain, texture, yield, moisture retention and shelf life; permits use of increased levels of inert ingredients (e.g. bran); flavour development
Kaiser (hard) rolls[1]	2.0% bakers	Improves structure, shelf-life, moisture retention; enhances appearance and eating quality
Hamburger buns[1]	2.0% bakers	Improves softness of bun dough; enhances appearance, freshness, shelf-life and machinability of dough
High-protein pasta	1.6% formula	Brings up protein level
Pizza crust	1.0 to 2.0% bakers	Supplements and strengthens the natural flour protein; reduces moisture transfer from sauce to crust; provides chewiness, strength and body
Bran bread[2]	2.0 parts by weight	Improves loaf volume, texture, yield, moisture retention and shelf-life; permits use of increased levels of inert ingredients; provides natural flavour enhancement
Brown soft rolls[2]	2.0 parts by weight	Improves softness of bun dough; enhances appearance, freshness, shelf-life and machinability of dough
Vienna bread[2]	2.0 parts by weight	Improves uniformity of loaf volume, grain and texture; gives longer shelf-life due to better moisture retention

Table 13.11 *Continued*

Baked product	Optimum level	Gluten function
Bread with low slice weight (high specific volume)[3]	30.0 parts by weight	High volume of vital wheat gluten is absolutely essential; also enhances overall appearance, keeping and eating qualities
Multi-grain bread[3]	5.0% bakers	Provides strength and stability to overcome degassing effect of sharp particles of bran or wheat; produces better loaf uniformity, increased volume, improved grain and fewer cripples
'Salad' rolls[3]	4.0% bakers	Provides required dough softness and crumb strength; improves overall appearance and keeping qualities
Wholemeal fibre-increased bread[3]	6.0 parts by weight	Improves loaf volume, grain texture, yield, moisture retention and shelf-life; permits increased level of inert ingredients; provides natural flavour enhancement
Whole wheat bread from flaked wheat[4]	10% bakers	Improves loaf volume, grain texture, yield, moisture retention and shelf-life; permits use of increased levels of inert ingredients; provides natural flavour enhancement

Formula sources:
[1]American Institute of Baking.
[2]Flour Milling and Baking Research Association, Chorleywood, UK.
[3]Bread Research Institute of Australia.
[4]Institute for Cereals, Flour and Bread TNO, The Netherlands.

Table 13.12 Chemical leavening agents

(a) Leavening acid applications

Leavening acid	Applications	Comments
Monocalcium phosphate monohydrate (MCP)	Pancake mixes, cookie mixes, angel food cakes, double-acting baking powders	Increased batter aeration, greater volume, improved texture, most of CO_2 released in mixer
Monocalcium phosphate, coated (Coated MCP)	Prepared cake mixes, self-rising flour and corn meal, mixes	A slowly soluble coating retards release of CO_2; gas pancake and waffle released during mixing, bench time and in oven
Sodium acid pyrophosphate (SAPP)	Various types of doughnut mixes, refrigerated canned biscuits, baker's baking powder, all prepared mixes	Slower acting; use in combination with others; improves tenderness and moisture of baked product

Table 13.12 *Continued*

(a) Leavening acid applications

Leavening acid	Applications	Comments
Sodium aluminium phosphate (SALP)	All types mixes, especially pancakes, biscuits and muffins	Slowest acting; use in combination with others; improves tenderness and moistness of baked product
Dicalcium phosphate dihydrate	Cake mixes	Slow acting, Ca ions for viscosity; CO_2 release only in oven
Sodium aluminium sulphate (SAS)	In combination with fast-acting leavening	Not satisfactory alone, too slow; may accelerate rancidity in flour-based mixes
Glucono-δ-lactone (GLD)	Canned bread, chemically leavened pizza	Slow acting, activated in oven

(b) Leavening acid combinations

Application (Acid)	Baking soda % (flour basis)	Reasons for use
Cake doughnuts (SAPP)	1.1–2.0	Little action while mixing, holding, extruding. Fast reaction in fryer.
Layer cakes (MCP, SAPP, SALP)	1.7–2.3	Nucleation. Late bake reserve. Al and Ca ions for batter viscosity and crumb resiliency.
Corn bread and muffins (MCP, Coated MCP, SALP)	1.5–2.0	No advantage for extra delay—poor gas retention.
Pancakes and waffles (MCP, Coated MCP, SALP)	2.0–2.5	Much nucleation needed. No reaction needed while holding, fast reaction on griddle.
Refrigerated biscuits and refrigerated doughs (MCP, Coated MCP, SALP)	2.0–2.3	Little reaction during processing. Some reaction after canning. Most reaction in oven.
Biscuits (MCP, Coated MCP, SALP)	1.4–2.0	Some delayed reaction. Fast oven action (short bake).

(c) Neutralising values

Crude monocalcium phosphate monohydrate	80–85
Sodium acid pyrophosphate	72
Monocalcium phosphate, monohydrate	80
Anhydrous coated monocalcium phosphate	83
Sodium aluminium phosphate, acidic, with aluminium sulphate, anhydrous	100

Table 13.12 *Continued*

(c) Neutralising values	
Sodium aluminium phosphate, acidic, with anhydrous, coated monocalcium phosphate	80
Sodium aluminium sulphate (SAS)	100
Potassium acid tartrate	45
Glucono-δ-lactone (GLD)	45
Dicalcium phosphate, dihydrate	33
Sodium aluminium phosphate, acidic	100

Further reading

Abrol *et al.* (1984) *Advances in Cereal Science and Technology*, Volumes 1–9.
American Institute of Baking (1987) Chemical leavening agents, *Bakers Digest* **12** 32.
American Institute of Baking; Manhattan, Kansas (1984) *Technical Bulletin.*
Anon (1979) Adjustment for water, *AIB Technical Bulletin.*
Anon (1987) Bakery additives and their function, *Bakers Digest* **12** 19.
Anon (1987) Properties of amylases for commercial applications, *Bakers Digest* **12** 24.
Boge, J. A. (1985) *ASBE.*
Cain, G. L. (1986) *ASBE.*
Cain, G. L. (1987) Approved oxidising and reducing agents, *Bakers Digest* **12** 36.
Gorton, L. *et al.* (1987) *Bakers Digest.*
Gorton, L. *et al.* (1988) *Bakers Digest.*
Jackel, S. S. (1987) *ASBE.*
Labaw, G. D. (1982) Chemical leavening agents and their use in bakery products, *Bakers Digest* **2** 16.
Langhans, R. K. (1971) *Bakers Digest.*
Nesetril, D. M. (1967) *Bakers Digest.*
Nesetril, D. M. (1987) Sweeteners for baking, *Bakers Digest* **12** 39.
Ponte, J. R. (1971) *Wheat Chemistry and Technology*, AACC.
Pyler, E. J. (1988) *Baking Science and Technology*, Volumes 1 and 2.
Schuster, G. (1984) *Advances in Cereal Science and Technology*, AACC.
Strietelmeier, D. M. (1988) *AIB Technical Bulletin.*

14 Gases

G. R. McCAIN

14.1 Introduction

Industrial gases have a wide range of uses and functions in foods and food processing. The atmosphere in which food is stored can affect its stability, colour, flavour and texture. Gases incorporated into products can play a major part in determining the texture of a product. Liquefied gases are widely used to chill and freeze food products.

Nitrogen and carbon dioxide are the two most widely used gases in the food industry. Oxygen, hydrogen and small quantities of argon, carbon monoxide, ethylene and other speciality gases have limited uses. Nitrogen, oxygen and argon are considered atmospheric gases and are obtained by separation of the individual gases from the air. Exclusive of water vapour, the approximate composition of atmospheric air is given in Table 14.1. There are, in addition, minor amounts of neon, helium, krypton, xenon, hydrogen, methane, and various other substances from plant, animal, industrial, and natural sources. The primary commercial process for air separation is the fractional distillation of liquefied air. Through fractional distillation high purity gases are obtained, generally in excess of 99.99% purity.

Systems based on pressure swing adsorption and membrane separations have come into significant use for on-site production of gaseous nitrogen or oxygen. These systems do not produce such high-purity gases as are obtained from the distillation systems, but do provide economical sources of nitrogen or oxygen for many applications. Some of these on site systems for nitrogen production can produce nitrogen with less than 0.1% oxygen. Using these separation techniques, argon does not separate from nitrogen, but since the argon is completely inert, this is not a problem for the normal applications of nitrogen in the food industry. These systems operate at or near ambient temperature and produce gaseous nitrogen or oxygen but cannot produce liquid nitrogen for chilling or freezing.

Carbon dioxide may come from naturally occurring sources or as a by-product of certain industrial processes. Large quantities of carbon dioxide have been found in underground structures, many of which are associated with natural gas or oil deposits. In these areas, wells can be bored and carbon dioxide can be pumped out of the ground and purified. A number of industrial processes have by-product streams that have a very high level of carbon dioxide. The carbon dioxide from these processes can be purified and liquefied.

The costs of industrial gases vary from one location to another and are dependent on the quantities being used. The liquefaction, separation, and purification are energy-intensive processes, therefore production costs will be affected by energy costs in the area where the gas is being separated and purified. The shipping distance between the gas-processing facility and the user will be a significant cost factor. In addition to the variable costs of energy and transportation, carbon dioxide has a cost for the raw carbon dioxide which varies and the potential locations for processing facilities are dependent on a suitable carbon dioxide source. Air separation plants that produce nitrogen and oxygen can select locations based on demand and energy costs in a given area. Because the factors that control the costs of carbon dioxide and nitrogen are different, the relative costs of the two vary from area to area.

Table 14.1 Composition of atmospheric air

Component	Formula	Percentage
Nitrogen	N_2	78.08
Oxygen	O_2	20.95
Argon	Ar	0.93
Carbon dioxide	CO_2	0.035

14.2 Freezing and chilling

The word *cryogenic* refers to very low temperatures. Both nitrogen and carbon dioxide are considered cryogens because of the low temperature of these liquids or solids at atmospheric pressure. These low temperatures are what make liquid nitrogen and liquid or solid carbon dioxide such efficient materials for freezing or chilling. When the low-temperature liquid or solid is brought in contact with the food to be chilled or frozen, a large amount of heat is removed from the product as the cryogen goes through a change of state. After the cryogen has contacted the food product it will completely change to the gas phase and there will be no residual carbon dioxide or nitrogen remaining in the product. Although nitrogen and carbon dioxide are both effective cryogens and are both widely used in food freezing, the two materials have some different characteristics which can be very important in some applications.

Liquid nitrogen boils and becomes a vapour at $-196°C$ ($-320°F$). In freezing applications, it can be used as a liquid where the product to be frozen is immersed in the liquid nitrogen at $-196°C$. Alternatively, the product can be frozen on a belt or in a batch freezer where it is exposed to a combination of cold nitrogen vapour and a spray of fine droplets of liquid nitrogen which immediately vapourise when they contact the product.

Carbon dioxide has the unique property of not existing as a liquid at atmospheric pressure. At atmospheric pressure, solid carbon dioxide (dry ice) exists at $-78°C$ ($-109°F$). As the temperature increases, the solid carbon dioxide sublimes and becomes a gas without going through a liquid phase. Liquid carbon dioxide will not exist below a pressure of 517.97 kPa (5.112 atm). If liquid carbon dioxide under pressure is released to atmospheric pressure, it will become a mixture of solid and gaseous carbon dioxide. As the liquid carbon dioxide is released to atmospheric pressure, some of the liquid will expand and bacome vapour. During the vapourisation, heat is absorbed from the surrounding liquid carbon dioxide and its temperature is reduced to the point where it becomes a solid. The initial temperature of the liquid will determine how much of the liquid becomes vapour and how much becomes solid. So, when liquid carbon dioxide is sprayed onto a product, as it is released to atmospheric pressure, some of it will form fine solid particles. As the fine particles contact the product, they will sublime and rapidly lower the temperature of the product.

In many freezing and chilling applications, carbon dioxide and nitrogen are both suitable and the selection will be dependent on the economics and availability of the two cryogens in the area. In other cases, the differences become more important and may favour one over the other. Detailed comparisons of cooling and freezing efficiencies of carbon dioxide and nitrogen are quite complex and are greatly influenced by the specific system being considered. Carbon dioxide has a higher latent heat (the heat required for a change of state from a solid or liquid to a gas) than nitrogen, but nitrogen has a higher sensible heat (the heat required to raise the temperature of a given weight of gas by unit temperature) than carbon dioxide. Because carbon dioxide does not exist as a liquid at atmospheric pressure, it cannot be used in an immersion freezer where the product is dropped into the liquid cryogen. Solid carbon dioxide as a snow is often preferred for cooling where the carbon dioxide snow can be blended into a food product in a mixer. For most applications, either liquid nitrogen or carbon dioxide can be used for chilling or freezing. Some equipment that is designed specifically for one cryogen cannot be used with the other because of the differences between the two.

In most applications where nitrogen or carbon dioxide is used for freezing and chilling, the cryogen will be delivered in bulk as a liquid and stored on site in a bulk tank. To understand the handling and storing of the different gases, you must consider the characteristics of each gas and understand the pressures and temperatures associated with them. Nitrogen and carbon dioxide have the physical properties shown in Table 14.2.

The critical point temperature is the highest temperature at which the gas will exist as a liquid regardless of pressure. Above the critical temperature, the nitrogen or carbon dioxide will always be a single phase, there cannot be a two-phase system with a liquid and vapour. Therefore, in order to store nitrogen as a liquid, it must be stored below its critical temperature of $-146.9°C$. Carbon dioxide may be held as a liquid as long as the temperature is below the critical point of $31.0°C$, providing the pressure is high enough.

Table 14.2 Physical properties of nitrogen and carbon dioxide

Physical property	N_2	CO_2
At 21.1°C and 101.3 kPa (1 atm):		
State	gas	gas
Specific volume ($m^3\,kg^{-1}$)	0.862	0.547
Density ($kg\,m^{-3}$)	1.160	1.828
Critical temperature (°C)	-146.9	31.0
Critical pressure (kPa)	3400	7381.5
Boiling point at 101.3 kPa (°C)	-195.8	$—$[1]

[1]Liquid does not exist at atmospheric pressure

Liquid nitrogen bulk storage tanks are vacuum-insulated tanks that hold the liquid at a low temperature and pressure. The tanks are generally rated at 1724 kPa (17 atm) maximum pressure and the pressure is controlled by a pressure regulator that will control to the desired pressure. If the pressure exceeds the set pressure, a small amount of gas will vent from the tank. When this happens, the pressure will drop and a small amount of liquid nitrogen will vaporise and reduce the temperature of the liquid nitrogen. By controlling the pressure, the temperature will be maintained at the temperature that has a vapour pressure equal to the set pressure. Nitrogen is often held at 690–1034 kPa, but, for freezing applications, it is often desirable to hold the nitrogen at the lowest pressure possible that will give satisfactory flow through the lines and spray from the nozzles. This can be as low as 170 kPa.

Carbon dioxide bulk storage tanks are foam-insulated tanks. Because of the difference between the gases, the storage conditions of carbon dioxide are different than the conditions used for nitrogen. Since liquid carbon dioxide does not exist below a pressure of 517.97 kPa, liquid carbon dioxide bulk tanks must be held at a pressure above 517.97 kPa. If the pressure were to drop below 517.97 kPa, the carbon dioxide would go to a solid and a vapour. Carbon dioxide tanks are normally rated at 3100 kPa, and the carbon dioxide is held at higher pressures than the liquid nitrogen. Carbon dioxide is generally held at between 1380 and 2070 kPa. It is commonly stored as a liquid at $-18°C$ which requires a pressure of 2007 kPa. The tanks for carbon dioxide often have a mechanical refrigeration system to maintain the temperature as opposed to the nitrogen tanks which maintain temperature by venting off excess pressure.

14.3 Types of freezing and chilling equipment

14.3.1 Immersion freezers

An immersion freezer consists of a tank that holds liquid nitrogen with a belt running through the liquid nitrogen. In the immersion freezer, the product to be frozen is placed on a wire belt and carried into the liquid nitrogen or dropped directly into the liquid nitrogen and then carried out the opposite end. The product normally has a very short residence time in the liquid nitrogen. Often the immersion freezer is followed by a tunnel to allow the product to come to temperature equilibrium. When the product exits from the liquid nitrogen bath, the outer surface of the product will be approaching the temperature of the liquid nitrogen $(-196°C)$ while the centre of the product is at a much higher temperature. Immersion freezing is the most rapid freezing method and it is especially well suited for products that are sticky and need to be very rapidly frozen so they will not stick together. Soft products that will change their shape if not frozen very rapidly often use immersion freezing. There are some products where there are large pieces and the large temperature differential associated with the immersion freezing may cause freezing stress and cracking. These products may be better handled by one of the other forms of freezers, such as a freezing tunnel. A variation on the typical immersion freezer is set up with an inclined auger which carries the frozen product out of the liquid nitrogen. This is used for freezing liquid products, such as eggs or cream. The liquid product can be fed dropwise into the liquid nitrogen and will freeze as a sphere when it hits the liquid nitrogen. It is then carried out of the liquid nitrogen by the auger.

14.3.2 Cabinet-type batch freezers

In a batch freezer, product is loaded onto racks and then wheeled into the cabinet where it goes through a freezing cycle. The freezer can operate with either liquid nitrogen or carbon dioxide. The cabinets contain fans that give good mixing and movement of the cold gases. By controlling the temperature of the freezer and the amount of cryogen that is sprayed into the freezer, the rate of freezing can be controlled. These are well suited for small freezing operations. In recent years, they are seeing a considerable usage with the cook-chill process for food service and 'deli' operations. Their flexibility makes them well suited for these operations, where rapid, well-controlled chilling is critical. For cook-chill applications, a cabinet freezer is normally equipped with a ramping-type controller. The cooling cycle will start at a rather low temperature to maximise cooling and gradually increase the temperature so that the product is chilled without freezing. The cooling cycle can be changed easily for different products. A large food service operation may well find that several cabinet units with programmable cooling cycles which allow different cycles to be run for different products serves their needs better than a continuous unit, such as a tunnel. Since a food service operation is likely to be preparing a number of different items that require different cooling cycles, it is beneficial to have several cabinet units so it is possible to have different cycles running simultaneously.

14.3.3 Freezing tunnels

A freezing tunnel is a belt running through a long tunnel. The tunnels are quite suitable for use with either nitrogen or carbon doxide. The tunnel has spray nozzles for the cryogen and fans that ensure rapid contact between the cryogen, the cold gas and the product. Tunnels can be designed and operated either isothermally with a uniform temperature throughout or countercurrent where the cryogen is introduced near the exit of the tunnel and the cold vapours flow back over the incoming product. Although generally used for freezing, by adjusting the temperature of the tunnel, they are a very effective means to cool warm product to refrigeration temperatures.

Freezing tunnels are not normally operated at the boiling point of the cryogen, rather at the temperature that will rapidly freeze the product with the most efficient operation. Tunnels operate over a wide range of temperatures. For freezing they normally operate from $-40°$ to $-100°C$. They can be run at temperatures above $-40°C$ for cooling applications. The cooling and freezing is accomplished by a combination of contact with cold vapours and contact with fine droplets or flakes of the cryogen.

14.3.4 Spiral freezer

The spiral freezer is a high-capacity, highly space-efficient type of freezer. The product is carried through the freezing chamber on a belt that follows a helical pattern. The product is feed into the bottom of the freezing chamber and makes a number of circular paths around the inside and finally exits the top of the freezer. The cryogen is sprayed into the freezer and moved around by fans so the spiral operates isothermally. Their compact design makes them very efficient to operate and minimises the floor space required. The spiral operates on either nitrogen or carbon doxide.

14.3.5 *Snow horns*

There are many processes in food processing that create unwanted heat in the product. In the meat-processing industry, grinding and mixing require that large amounts of energy be put into the products. Much of this energy is converted to heat. In most meat products, temperature control is critical and the product must be held in a very narrow temperature range during processing. During the grinding and mixing the temperature will rise. If it is not controlled, the temperature increase will cause problems such as fat-smearing or the breaking of emulsions. The use of cryogens gives a simple and quick way to remove heat from heavy products that are being mixed. The most common cryogen for this application is carbon dioxide snow. Blenders for meat mixes are routinely fitted with carbon dioxide snow horns. These snow horns are fed with liquid carbon dioxide which becomes a fine snow when it hits atmospheric pressure. The snow will rapidly mix with and cool the product. As it cools the product the carbon dioxide will vapourise so that it does not add anything to the product or change its composition. Doughs in the baking industry are very temperature-sensitive also. Carbon dioxide snow can be used to cool the dough to maintain the proper temperature. Carbon dioxide snow is regularly used to chill poultry and other highly perishable products as they are packed. Carbon dioxide snow can be placed in boxes with the product without the problems or mess that arises when these products are packed with ice which melts into water.

14.3.6 *Ingredient chilling*

In hot weather, flour and other ingredients are often warm going into the mixer which creates product problems. Cryogens are generally the most efficient way to cool flour and other dry ingredients. Several systems have been developed for injecting cryogens into augers and pneumatic conveying systems to cool dry ingredients.

14.4 Gas atmospheres

Foods can go through a variety of changes during storage including oxidation, enzymic changes, and microbial growth. One of the primary functions of food processing is to get the food product to the end consumer in acceptable condition. Controlling the gaseous environment in which the product is stored is an important tool available to the food technologist to accomplish this goal. Altered gaseous atmospheres fall into two broad categories, controlled atmosphere and modified atmosphere or modified atmosphere packaging (MAP). MAP refers to the process where the normal air in a package is replaced by a specific gas or blend of gases. Once the package is sealed, the atmosphere may change as a result of actions of the product, bacterial action, and gas permeation through the film. Controlled atmospheres refer to systems where a specific atmosphere is created and then controlled on an ongoing basis throughout the storage period.

Fruits and vegetables, during the time they are growing, are undergoing photosynthesis where they absorb carbon dioxide and give off oxygen. Once the fruit or vegetable is harvested, it is no longer growing, however it does continue to respire. The harvested fruit or vegetable goes into aerobic respiration and absorbs oxygen and starts to break down cell components and give off carbon dioxide. If the respiration of the fruit or vegetable can be slowed, it can greatly extend the acceptable life of the product. A good example of the use of controlled atmosphere is the commercial storage of apples. With refrigerated storage in an atmosphere of air, apples only last a few weeks. Initially, apple storage facilities used only low temperature to store the apples. By putting the apples in a sealed, refrigerated storage facility, the respiring apples would soon use up the available oxygen and build up significant levels of carbon dioxide in the atmosphere. The depletion of oxygen did help extend the storage life of the apples. However, there were problems caused by the lack of oxygen and the build up of ethylene. It is now established that apples and most other fruits and vegetables are best held in an atmosphere that contains a low level of oxygen. The desired carbon dioxide level for different products varies widely. Most apples are now stored in a controlled atmosphere where the atmosphere is constantly circulated and controlled. The ideal atmosphere for apples is generally between 1 and 3% oxygen and 1 and 8% carbon dioxide with a high relative humidity and no ethylene. The on-site pressure swing adsorption and membrane separation systems are very well suited for this application and are becoming widely used for controlled atmosphere storage. With these systems, the atmosphere can be precisely controlled and extended storage times can be achieved.

A complicating factor with storage of fruits and vegetables, is that each product has its own set of ideal conditions. Even within a single fruit or vegetable, there are significant differences between cultivars as to the ideal temperature, relative humidity and gas composition. Almost all atmospheres recommended for fresh fruits and vegetables contain a reduced level of oxygen, some carbon dioxide (up to about 20% maximum), with the balance being nitrogen. Each product has a tolerance level for low oxygen and high carbon dioxide. Oxygen levels below the minimum or carbon dioxide above the maximum will cause damage to the product.

Some work has been done using carbon monoxide in controlled or modified atmospheres. Carbon monoxide, unlike the other gases used in modified atmospheres, is very toxic and can be extremely hazardous in the workplace.

A large amount of information has been published on the effects of different atmospheres on different products. Some examples of gas mixtures for different products are listed in Table 14.3. The examples list the percentage of carbon dioxide and oxygen, with the balance being nitrogen.

Table 14.3 Examples of modified atmospheres for food packaging

Food	Carbon dioxide	Oxygen
Nuts	0	0.5
Celery	1–2	2–3
Pears	1–2	2–3
Citrus fruits	0–10	5+
Apples	1–6	1–3
Broccoli	5–10	2–3
Peppers	5–10	3–4
Asparagus	10	5–10
Sweet corn	15	2–3
Cherries	10–15	2–3
Strawberries	20	2–3

Meat, poultry and fish present a different set of problems and conditions. The primary cause of spoilage and degradation in muscle foods is bacterial growth with enzymic degradation also being an important factor, especially in fish. In order to store any muscle food more than a few weeks, it must be preserved in some manner such as freezing, canning or drying. Traditionally, the majority of these products are processed and sold fresh and are held refrigerated through distribution and sale. Even with good refrigeration and sanitation, the shelf-life of these products is relatively short. It has been demonstrated that packaging these products in a modified atmosphere can significantly increase the shelf-life. Carbon dioxide has a strong inhibitory action on many of the normal spoilage organisms and also inhibits some enzymic degradation. An atmosphere that is high in carbon dioxide and low in oxygen is effective in increasing the shelf-life of these items (Table 14.4). The use of modified atmosphere to increase the shelf-life of these products has raised some concerns that have slowed the use of this technology, especially in North America. These products can be contaminated with a number of pathogenic organisms. With normal atmospheric packaging, these pathogenic organisms either do not grow at the storage temperatures at which these products are held or, if they do, they grow slowly and cannot compete with the normal spoilage organisms. In a modified atmosphere, the normal spoilage organisms are inhibited and the shelf-life is extended. This

Table 14.4 Examples of gas mixtures used with muscle foods

Food	CO_2	O_2
Red meat	30–50	50–80
Poultry	30–60	1–5
Fish (oily)	40–110	0–2
Fish (non-oily)	40–100	0–40
Crustaceans	80–100	1

presents the possibility of pathogenic organisms growing in an environment free of competitive organisms. Many of these pathogenic organisms do not produce the normal indications of spoilage such as off-odours and slime. Of special concern is the growth of the non-proteolytic *Clostridium botulinum*, some of which can grow at temperatures as low as 4°C. Since the generally recommended atmospheres for seafood and poultry are high in carbon dioxide with little or no oxygen, these environments could allow the growth and toxin production from *C. botulinum* if temperatures are not very closely controlled. *Listeria monocytogenes* will grow at very low temperatures and can grow with little or no oxygen.

The red meats are somewhat different than the fish and poultry. The consumer associates a bright red colour with freshness and quality. If meat is packaged in a carbon dioxide environment, it will have a reddish-purple colour and not a bright red colour. The bright red colour in meat is caused by oxygen reacting with the myoglobin and creating oxymyoglobin, which is the bright red pigment of meat. In order to have packaged meat with bright colour and extended shelf-life, a modified atmosphere that is very high in oxygen and carbon dioxide is used. The oxygen will give a bright red colour and the carbon dioxide will inhibit spoilage bacteria and extend the shelf-life.

In bakery products, the problems are mould growth and staling. An atmosphere high in carbon dioxide and free of oxygen will prevent mould growth and retard staling in some products. The atmosphere should not be too high in carbon dioxide, because the product will absorb the carbon dioxide. If there is not adequate nitrogen in the atmosphere, as the carbon dioxide is absorbed, a vacuum will be created in the package and can collapse the package and crush the product. In some products, such as nuts and snack foods, the primary deterioration is the rancidity of the oil. By packaging the product in an atmosphere that is free of oxygen, generally pure nitrogen, rancidity can be retarded.

Modified and controlled atmospheres are generally used with other factors, such as refrigeration, to extend the shelf-life of products. The main gases used are carbon dioxide, nitrogen and oxygen. Carbon dioxide will inhibit the growth of a wide range of bacteria and will penetrate and thus inhibit some enzymic degradation in products. Nitrogen is used where an inert gas is needed. Oxygen is used for those products that will develop off-flavours or other problems when subjected to a low-oxygen or oxygen-free atmosphere. When products are packaged with a modified atmosphere the packaging materials must be carefully selected since different films have different permeabilities for the different gases. If a proper film is not selected, the desired atmosphere may escape through the package in a very short period of time.

14.4.1 *Inerting*

There is a number of products where oxidation can result in rapid product degradation. Certain flavours are very susceptible to oxidation, such as some of the flavour components of orange. For this reason it is common to store and pack orange juice with a nitrogen atmosphere.

Fats with unsaturated fatty acids will oxidise and develop off-flavours. Many snack items are fried in oil or have a high oil content. These products are often packed in nitrogen to prevent oxygen from contacting the product. In the processing and storage of oils, the tanks are maintained with a headspace of nitrogen to protect the oil from oxygen.

During the processing of oxygen-sensitive products, the products may be sparged with nitrogen to remove dissolved oxygen. Both nitrogen and oxygen have rather low solubilities in water, but some gas will be dissolved. Given adequate time, the gases dissolved in a liquid will come to equilibrium with the environment around it. If the environment is air, nitrogen and oxygen will be dissolved in the liquid. The amount of each gas is dependent on the fraction of that gas in the contacting atmosphere and the solubility of that gas in the liquid.

The solubilities of gases in liquids are often expressed as millilitres (ml) of gas per millilitre of liquid. The solubilities of gases in liquids are generally quite temperature dependent with the solubilities decreasing as the temperatures increase. An example of the solubility of nitrogen and oxygen in water is shown in Table 14.5 at 25 and 50°C.

The data in the table show the solubility of oxygen and nitrogen when water is in contact with pure oxygen or pure nitrogen and water is in contact with air. As can be seen from the figures, the solubilities decrease with

Table 14.5 Gas solubility in water (ml gas (ml H_2O)$^{-1}$)

Temperature (°C)	Oxygen (O_2)	Nitrogen (N_2)	Air Total	N_2	O_2
25	0.0283	0.0140	0.0167	0.0110	0.0058
50	0.0209	0.0108	0.0114	0.0075	0.0039

increasing temperature. When exposed to air, the amount of dissolved nitrogen and oxygen is approximately equal to the fraction of the gas in the air times its solubility. These relationships are the basis for sparging. Given a liquid that has come to equilibrium with air, it will contain dissolved oxygen in proportion to the oxygen content of the air and its solubility in the product. If a product is sparged with nitrogen and nitrogen gas is bubbled through and mixed with the liquid, an equilibrium with the gas will be reached. Assuming the volume of the nitrogen used is greatly in excess of the amount of dissolved gases, the liquid will come to equilibrium with the gas phase in the bubbles that contain almost no oxygen and almost all the oxygen will be removed from the liquid. After sparging, the liquid will be saturated with nitrogen. For this reason, it is common to sparge oxygen-sensitive products with nitrogen prior to storage. After the sparging, the product will be held under nitrogen to prevent oxygen from re-entering the product.

Understanding these relationships, it becomes apparent that in systems where oxygen is needed, it is more effective to sparge with pure oxygen than with air. Examples of where oxygen is needed are waste-water treatment where there are high oxygen demands to break down organic matter and intensive aquaculture systems where a high concentration of fish creates a high oxygen demand. It is also apparent that when the waste temperatures are higher, the sparging will give a lower amount of dissolved oxygen. This is unfortunate, because the growth of bacteria in waste treatment or fish in aquaculture is generally more rapid at higher temperatures and creates a higher oxygen demand.

14.5 Special uses

14.5.1 Carbonated beverages

Carbon dioxide has a long history of use with carbonated soft drinks. The carbon dioxide is responsible for the 'fizz' and some of the sharp flavour characteristics of these beverages. Carbonated beverages are based on supersaturated solutions of carbon dioxide. These products are carbonated under pressure, generally about 345 kPa, so that about four times as much carbon dioxide is dissolved as would be dissolved at atmospheric pressure. When the beverage is poured and consumed, some of the carbon dioxide will come out of solution and create the fizz and foam associated with these beverages.

There are several differences between carbon dioxide and nitrogen when they are dissolved in water. Some solubility figures for carbon dioxide in water are listed in Table 14.6. A comparison of these figures with the

Table 14.6 Solubility of CO_2 in water (ml ml^{-1})

Temperature (°C)	Pressure (kPa)			
	101	207	310	413
0	1.7	3.46	5.21	6.95
25	0.83	1.53	2.30	3.07
32	0.71	1.27	1.91	2.56

previously given data for nitrogen shows that carbon dioxide is roughly 60 times as soluble in water as nitrogen.

When nitrogen dissolves in water it acts as an inert gas and does not react with the water. A certain amount of carbon dioxide in solution will react with the water to form carbonic acid and can lower the pH of the water:

$$H_2O + CO_2 \rightleftharpoons H_2CO_3 \qquad K = 2.3 \times 10^{-3}$$

$$H_2CO_3 \rightleftharpoons H^+ + HCO_3^- \qquad K = 1.7 \times 10^{-4}$$

A saturated solution of carbon dioxide in water at atmospheric pressure will have a pH of approximately 4.2. Most flavoured, carbonated soft drinks have other acidic ingredients and have a pH far below 4.0, so that carbon dioxide does not really contribute to the acidity of the product.

A situation where the carbon dioxide can be an important factor is the pH in carbonated bottled waters. The use of bottled water in North America has increased greatly in the last few years. It has been shown that there is concern over the fact that a few species of bacteria, primarily *Pseudomonas*, will grow on the micro-nutrients they can get from the water and plastic containers. Carbonation of the water at increased pressure appears to prevent the growth of *Pseudomonas aeruginosa*, the main organism of concern.

Non-carbonated beverages, such as juice drinks or iced teas packed in thin-walled containers, are often packed with nitrogen to pressurise the container so that the container will hold its shape and not collapse when cases are stacked. Two different methods have been used to accomplish this. One method uses droplets of liquid nitrogen. A droplet of liquid nitrogen is added to the container just before it is sealed. The vaporising liquid will build adequate pressure in the container to rigidify it. Since this type of product is often packed in plants that are packing carbonated beverages, a non-carbonated beverage can be saturated with nitrogen at the pressures used for carbonation. After the container is closed, the nitrogen will come to an equilibrium pressure with the liquid and give some pressure to the container. The pressure is much less than the pressure achieved with a carbonated beverage. In some cases argon has been used instead of nitrogen for this application. Argon is totally inert and has never been shown to form any stable compounds. Argon has about twice the solubility of nitrogen, therefore a saturated solution of a liquid at an elevated pressure will have about twice the volume of dissolved argon as a comparable system with nitrogen. When the argon equilibrates with the gases in the container it will yield a higher internal pressure.

14.5.2 Supercritical fluid extraction

As has been described previously, carbon dioxide has a critical temperature of 31.0°C and a critical pressure of 7381.5 kPa. Above the critical pressure and temperature, carbon dioxide exists only as a single phase, a supercritical fluid. As the pressure increases, the density increases, but the fluid stays as a single-phase fluid. It is well established that supercritical carbon dioxide is an excellent solvent for many fat-soluble compounds. Because of the many consumer concerns over organic solvents used for the extraction of flavours, caffeine and certain other compounds from foods, there is a considerable amount of interest in supercritical fluid extraction with carbon dioxide. The high pressures required for this process make the equipment expensive, but there are certain applications where supercritical extraction is coming into commercial use. Although the capital costs associated with supercritical extraction are high, the operating costs are relatively low. In many applications there is a major benefit in replacing an organic solvent that is expensive and hazardous with carbon dioxide. There are plants operating for the production of hop extracts for the brewing industry, decaffeination of coffee and tea and extraction of flavour compounds from spices. Another application that is currently under development is the removal of cholesterol from food products such as butter or eggs.

In supercritical fluid extraction, supercritical carbon dioxide is pumped through the product and will extract certain compounds. After the extraction, by increasing the temperature or decreasing the pressure, the density of the carbon dioxide will decrease and the extracted material can be separated from the carbon dioxide. Supercritical fluid extraction is a complex technology and the process is expensive, however there are specialised applications in the food industry where it is being adopted.

14.5.3 Carbon dioxide fumigation

A considerable amount of work was done in the 1950s and 1960s using carbon dioxide as a fumigant for grains and other food products. At that time there were a number of chemical fumigants available that were more economical and effective than carbon dioxide, so the carbon dioxide was never developed as a fumigant. During the last few years, a number of previously used fumigants have been banned and regulations covering the remaining fumigants have become much more restrictive. There are major consumer concerns about the use of all pesticides and fumigants. This has prompted a re-evaluation of carbon dioxide as a fumigant. Both carbon dioxide and nitrogen have potential as a fumigant. Nitrogen is strictly an asphyxiant and in many cases the oxygen content must be reduced to below 2% to be effective. Carbon dioxide is generally effective at concentrations of 50–60% and is not dependent on producing a low oxygen content. Since many bulk storage structures are not tightly sealed, it is more practical to establish and maintain a 60% concentration of carbon dioxide than to establish 2% oxygen with nitrogen. Treatment with carbon dioxide or nitrogen will effectively kill all insect infestation in 2–4 days with the above levels.

14.5.4 Ozone

Ozone is probably most widely known for the ozone layer in the earth's atmosphere or as an undesirable compound that forms from the action of sunlight with certain pollutants in the air. Ozone has a number of potential uses in the food industry because of its strong oxidising power and bactericidal action. It has been used as a bactericidal agent in water treatment. It has been more widely used in water treatment in Europe then in North America, but its use in North America is increasing, primarily as a result of safety concerns of chlorine. Chlorine has been shown to react with some organic compounds that may be found in water and form chlorinated compounds that are carcinogenic. Ozone appears to be a potential replacement for chlorine in many applications.

Ozone is not sold directly, but is generated on site with an ozone generator that converts O_2 oxygen to O_3 ozone. Most commercial ozone generators use a corona discharge to produce ozone. If the generator is fed with air, the air must be clean, oil free and dry ($-60°C$ dew point). If the air is not adequately dried there can be a problem with the production of nitrogen oxides and reduced efficiency. An oxygen feed to an ozone generator gives a much higher efficiency than an air feed and thereby reduces power costs. The oxygen feed will also give a higher ozone output than an air feed.

Ozone is a very strong oxidising agent that will destroy many objectionable odours, flavours, and harmful compounds. This makes it very useful in waste treatment. Ozone is being proposed to sterilise recycled poultry chilling water.

Ozone degrades to oxygen so it does not leave a residue. In dry air it has a half-life of 12 hours. In distilled water the half-life is 20–30 minutes. In water with an organic load, the half-life will be much shorter. It can be easily broken down in either air or water so that it will not escape to the surrounding environment. Like any strong oxidising agent or bactericidal agent, proper precautions must be taken when ozone is used.

14.5.5 Safety

Industrial gases are generally non-toxic and present relatively few hazards, but common sense and reasonable care need to be exercised. Oxygen in concentrations above atmospheric levels can promote rapid burning and at high concentrations it may react violently with certain materials.

Nitrogen is not normally harmful. We regularly breathe air that is 78% nitrogen. It can be an asphyxiant if the oxygen level becomes depleted.

Carbon dioxide is normally present in the air at 350 ppm. Exposure limits for carbon dioxide have been set at an average daily exposure limit of 5000 ppm (0.5%) and a short-term exposure limit of 30 000 ppm (3%) for 15 minutes. At this level, there will be a slight irritation to the eyes and breathing passages so that one will be aware of its presence and can increase the ventilation to the area or move to another area.

Liquid nitrogen or solid dry ice can cause 'freeze burns' if in contact with the skin. Cryogens are normally piped directly to the equipment in which they are being used, and the equipment is closed. Common sense and reasonable care should be used when working at cryogenic temperatures.

15 Chelating agents

T. NAUTA

15.1 Introduction

Free metal ions in food systems may form insoluble or coloured compounds or catalyse degradation of food components, resulting in precipitation, discoloration, rancidity or loss of nutritional quality.

Chelating agents eliminate these undesirable effects by forming stable, usually water-soluble complexes with free metal ions. This effect is called *chelation* and the complexes formed are referred to as *chelates*.

Another function of chelating agents is the controlled release of metal ions for nutritional purposes or for controlled gelation of thickeners. An overview of chelating agents is given in Table 15.1.

Chelation is an equilibrium reaction. The ratio of chelated to unchelated metal ion is indicated by the stability constant K. The higher the affinity of the chelating agent for a particular metal ion, the higher the stability constant will be. In Table 15.2 the stability constants of some metal chelates are given. The chelating value indicates the amount of metal ion chelated by 1 g of chelating agent. Some values are given in Table 15.3. Table 15.4 lists some properties of chelating agents.

The applications of chelating agents in foods are given in Table 15.5. Note that most EDTA doses are maximum permitted levels by the Food and Drug Administration. The use of particular chelating agents is not approved in all countries.

Table 15.1 Grouping of chelating agents in foods

Type	Example
Amino acids	Glycine
Aminocarboxylates	EDTA
Hydroxycarboxylates	Citric acid
	Gluconic acid
	Tartaric acid
Polyphosphates	Hexametaphosphoric acid
	Pyrophosphoric acid
	Tripolyphosphoric acid
	Phytic acid

Table 15.2 Stability constants of metal chelates (log K values)

Chelating agents	Ca^{2+}	Mg^{2+}	Fe^{3+}	Mn^{2+}	Cu^{2+}	Zn^{2+}
EDTA	10.7	8.8	25.1	13.9	18.8	16.5
Pyrophosphoric acid	5.4	5.5	—	—	7.6	8.7
Tripolyphosphoric acid	5.2	5.8	—	7.2	8.3	7.5
Citric acid	3.5	3.4	11.5	4.2	5.9	5.0
Tartaric acid	1.8	1.4	6.5	2.5	3.4	2.2
Glycine	1.4	2.2	10.0	2.8	8.2	5.0

Table 15.3 Chelating values (mg metal (g chelating agent)$^{-1}$)

Chelating agent	Ca^{2+}	Mg^{2+}	Fe^{3+}	Mn^{2+}	Cu^{2+}	Zn^{2+}
EDTA–CaNa$_2$	—	—	135	130	150	155
EDTA–Na$_2$H$_2$	105	65	150	145	70	175
Citric acid	205	125	290	285	330	340
Sodium pyrophosphate	150	90	205	205	235	240
Sodium tripolyphosphate	105	65	150	145	170	175

Table 15.4 Chemical and physical properties of some chelating agents

Product	Molecular weight	Physical form	Solubility g $(100 ml H_2O)^{-1}$	pH 1% in water	Specific points
EDTA–CaNa$_2$. 2H$_2$O	410.3	White powder	40	7	
EDTA–Na$_2$H$_2$. 2H$_2$O	372.2	White powder	10	4.5	
Citric acid	192.1	Colourless crystals	160	2	
Potassium citrate . H$_2$O	324.4	White crystals	167	9	Deliquescent
Sodium citrate . 2H$_2$O	294.1	White crystals or powder	71	9	
Glucono-δ-lactone	178.1	White powder	59	3	
Potassium gluconate	234.3	White powder	100	7	
Sodium gluconate	218.1	White crystals	60	7	
Tartaric acid	150.1	Colourless crystals	147	2	
Potassium sodium tartrate . 4H$_2$O	282.2	Colourless crystals	100	10	Efflorescent
Sodium tartrate . 2H$_2$O	230.1	Colourless crystals	33	10	
Tetrapotassium pyrophosphate	330.3	Colourless solid	187	10.5	Deliquescent
Disodium pyrophosphate	221.9	White crystals	15	8	(At elevated temperatures
Tetrasodium pyrophosphate	265.9	White crystals	5	10	(>100°C) condensed phosphates
Sodium tripolyphosphate	367.9	White powder	73	10	hydrolyse to less effective
Sodium hexametaphosphate	611.2	Colourless glass	VS[1]	10	chelating agents)

[1]VS – Very soluble.

Table 15.5 Applications of chelating agents

Food	Amino acids	Aminocarboxylates EDTA-Ca (mg kg⁻¹)	EDTA-Na2 (mg kg⁻¹)	Hydroxycarboxylates Citrate %	Gluconate %	Tartrate %	Polyphosphates Hexameta- %	Pyro- %	Tripoly- %	Phytate %	Function
Beverages											
Canned carbonated		33		0.1–1.0	*	*					Promote flavour retention
Distilled alcoholic		25									Promote colour and flavour retention and product clarity
Fermented malt		25	150								Antiguishing agent
Multivitamin preparation											With iron salts as a stabiliser for vitamin B_{12}
Cereals											
Frozen cooked rice							0.02				Promote texture retention
Ready-to-eat products containing dried bananas			315								Promote texture retention
Dairy products											
Cheese whey			1:1 with Ca	1.25:1 with Ca 3(Na-)							Flux improvement
Process cheese				0.3			*				Prevent fat separation
Skim milk			3000							0.3	Prevent clotting
Fats and oils											
Dressings, non-standardised			75								Preservative
French dressing			75	0.01							Preservative
Lard		75	75								Antioxidant
Mayonnaise		75									Preservative
Margarine		75		*(Iso-propyl-, stearyl-)							Antioxidant
Salad dressing		75									Preservative
Sandwich spread		100	100								Preservative
Sauces		75	75								Preservative
Vegetable oil				0.01							Preservative
Fish											
Clams (cooked canned)		340									Promote colour retention

Product	Additive							Function
Crabmeat (cooked canned)		275	0.25					Promote colour retention and retard struvite formation
Frozen fish			0.25					Promote colour retention and prevent rancidity
Gefilte fish balls or patties			50					Inhibit discoloration
Roe		250					*	Improve colour and flavour
Shrimps								Promote colour retention and retard struvite formation
Fruit								
Apple slices			100	1 (in peeling lye)		2 (in peeling lye)		Prevent browning
Bananas			4000					Prevent discoloration
Canned strawberry pie filling			500					Promote colour retention
Frozen fruit				0.1–0.3				Promote colour and flavour retention
Mandarin						0.4 (in wash water)		Facilitate membrane removal
Pecan pie filling			100					Promote colour retention
Spreads, artificially coloured and lemon- or orange flavoured			100					Promote colour retention
Meat								
Cooked beef			500 *		0.5	1		Antibacterial
Cooked chicken			*			*		Improve flavour
Fresh pork				0.01–0.1		0.1–2.0		Promote colour and flavour retention
Nitrite-free curing	Glycine		36	0.5–1.5[1]		0.3–1.0		Curing accelerator with ascorbic acid
Mineral supplements K, Ca, Mg, Fe, Mn, Cu, Zn	Glycine aspartate		*	*		*		Mineral carrier
Sweeteners								
Caramel						1:1 with Ca	1:1 with Ca	Promote texture retention
Non-nutritive			1000		*			Sequestrant
Thickeners								
Alginate	Glycine		0.07					Improve gelation
Starch			*	*				Improve gelation
Pectin			*	*				Improve gelation

Table 15.5 *Continued*

Food	Amino acids	Aminocarboxylates		Hydroxycarboxylates			Polyphosphates				Function
		EDTA-Ca (mg kg^{-1})	EDTA-Na2 (mg kg^{-1})	Citrate %	Gluconate %	Tartrate %	Hexameta- %	Pyro- %	Tripoly- %	Phytate %	
Vegetables											
Canned black-eyed peas			145								Promote colour retention
Canned cooked chickpeas			165								Promote colour retention
Canned kidney beans			165								Preservative
Canned white potatoes		110									Promote colour retention
Dried lima beans (cooked canned)		310			0.2[1]						Promote colour retention
Frozen white potatoes											Promote colour retention
Mushrooms (cooked canned)		200	100								Promote colour retention
Pickled cabbage		220									Promote colour, flavour and texture retention
Pickled cucumbers		220									Preservative
Potato salad		100									Preservative
Processed dry pinto beans		800									Promote colour retention
Miscellaneous											
Egg product		200		0.17							Preservative
Instant coffee									*	<0.2	Prevent foaming and scum formation
Spice extractives in soluble carrier		60									Promote colour and flavour retention

[1] As glucono-δ-lactone.

Further reading

Ablett, R. F. and Gould, S. P. (1986) Frozen storage performance of cooked cultivated mussels (*Mytilus edulis* L.). Influence of ascorbic acid and chelating agents. *Journal of Food Science* **51** (5), 1118–1121.

Furia, T. E. (1964) EDTA in foods. A technical review. *Food Technology* **18** (12), 50–58.

Hole, M. (1980) Chelation reactions of significance to the food industry. *Process Biochemistry* **15** (2), 16–19, 24.

Kilara, A., Witowski, M., McCord, J., Beelman, R. and Kuhn, G. (1984) Development of acidification processing technology to improve colour and reduce thermophilic spoilage of canned mushrooms. *Journal of Food Processing and Preservation* **8** (1), 1–14.

Kleyn, J. G. (1971) *Biological Preservation of Beer*. US Patent No. 4,299,853.

Schuster, G. (1981) Herstellung und Stabilisierung von Lebensmittelemulsionen. *Seifen, Oele, Fette, Wächse* **107** (14), 391–401.

Shah, B. G. (1981) Chelating agents and bioavailability of minerals. *Nutrition Research* **1** (6), 617–622.

Index